SpringerBriefs in Pharmaceutical Science & Drug Development

For further volumes:
http://www.springer.com/series/10224

Sikha Mandal • Jnanendra Rath

Extremophilic Cyanobacteria For Novel Drug Development

 Springer

Sikha Mandal
School of Pharmacy
University of Mississippi
Oxford, MS, USA

Research Associate
Department of Botany
Institute of Science
Visva Bharati (A Central University)
Santiniketan, West Bengal, India

Jnanendra Rath
School of Pharmacy
University of Mississippi
Oxford, MS, USA

Department of Botany
Institute of Science
Visva Bharati (A Central University)
Santiniketan, West Bengal, India

ISSN 1864-8118   ISSN 1864-8126 (electronic)
SpringerBriefs in Pharmaceutical Science & Drug Development
ISBN 978-3-319-12008-9   ISBN 978-3-319-12009-6 (eBook)
DOI 10.1007/978-3-319-12009-6

Library of Congress Control Number: 2014955249

Springer Cham Heidelberg New York Dordrecht London
© Springer International Publishing Switzerland 2015
This work is subject to copyright. All rights are reserved by the Publisher, whether the whole or part
of the material is concerned, specifically the rights of translation, reprinting, reuse of illustrations,
recitation, broadcasting, reproduction on microfilms or in any other physical way, and transmission
or information storage and retrieval, electronic adaptation, computer software, or by similar or
dissimilar methodology now known or hereafter developed.
The use of general descriptive names, registered names, trademarks, service marks, etc. in this
publication does not imply, even in the absence of a specific statement, that such names are exempt
from the relevant protective laws and regulations and therefore free for general use.
The publisher, the authors and the editors are safe to assume that the advice and information in this
book are believed to be true and accurate at the date of publication. Neither the publisher nor the
authors or the editors give a warranty, express or implied, with respect to the material contained
herein or for any errors or omissions that may have been made.

Printed on acid-free paper

Springer is part of Springer Science+Business Media (www.springer.com)

Contents

About the Authors

Dr. Jnanendra Rath was awarded a Ph.D. in Botany from Utkal University, Bhubaneswar, India, for his research on ecophysiological studies on the algae of Chilika Lake on the east coast of India. He served as an assistant professor in Visva-Bharati University, Santiniketan, India, and started his independent research on bioprospecting of cyanobacteria from extreme habitats. He established the Visva-Bharati Culture Collection of Algae (VBCCA) and is affiliated with the World Federation for Culture Collections (WFCC). His major research interests include ecophysiology and survival strategies of cyanobacteria in extreme environments as well as obtaining bioactive secondary metabolites from cyanobacteria. He is an alumnus of the United Nations Institute for Training and Research (UNITAR) and a partner of both the Expert Center for Taxonomic Identification and the United Nations Educational, Scientific and Cultural Organization (UNESCO). Dr. Rath is also a member of the Society of Wetland Scientists as well as several scientific bodies including the Phycological Society of America, American Association for the Advancement of Science, the Association of Microbiologists of India, and the Indian Phycological Society. He is the recipient of the Raman Fellowship under the Indo-US Twenty-First Century Knowledge Initiative and is currently a visiting scholar in the University of Mississippi School of Pharmacy, Oxford, Mississippi, USA. He has completed three externally funded research projects and has authored two books and 24 papers published in internationally recognized journals.

Dr. Sikha Mandal completed her M.Phil. in environmental science and was awarded a Ph.D. in Botany from Visva-Bharati University, Santiniketan, India, for her research on extremophilic cyanobacteria of eastern India. She worked as a senior research fellow and research associate in a project sponsored by the Council of Scientific & Industrial Research (CSIR). Her major research interests include drug discovery and bioprospecting of extremophilic cyanobacteria. Dr. Mandal was awarded a Federation of European Microbiological Societies (FEMS) Young Scientist grant and a Society of Wetland Scientists (SWS) Women Scientists award. She is currently a member of SWS and the Psychological Society of America. She has published four papers in internationally reputed journals and is the author of three book chapters.

Chapter 1
Introduction

1.1 Origin of Cyanobacteria and Evolution of Life

Cyanobacteria are gram-negative oxygenic photosynthetic organisms. Although invisible to the naked eye, cyanobacteria are an essential component of the earth's biota. They catalyze unique and requisite transformations in the biogeochemical cycles of the biosphere, produce oxygen, an important component of the earth's atmosphere, and stand for a large portion of life's genetic diversity. Iron, the key metal of the universe played an important part in the origin of cyanobacteria and further evolution of life. Iron 60 provides the evidence in favor of a supernova explosion being the trigger for the formation of the solar system. The survival of life on the planet is feasible due to the liquid iron of the earth's core that created the magnetic field, without which the early atmosphere would have been stripped away and life could never have evolved. Fe^{2+} -mediated RNA folding and catalysis, in combination with the paleogeological models, suggest that RNA functioned and evolved in association with Fe^{2+}. Evidence also shows that the photochemical redox-cycling of Fe^{2+}/Fe^{3+} could have provided the necessary high energy, reducing equivalents for the formation of H_2, CH_4, and other important ingredients for the assembly of amino acids in this abiotic world (Braterman et al. 1983; Huber and Wächtershäuser 1997).

Life on earth originated at least 3500 Ma and the principal mode of metabolism is thought to have been anaerobic fermentation of organic compounds, which had formed, abiologically in the primeval environment (Lemmon 1970). The ferredoxins (iron–sulfur protein) probably played a significant role in the development of fermentative bacteria. It is generally considered (Schilling 1973) that sulfur reduction was one of the earliest forms of microbial respiration, because the known microorganisms that are most closely related to the last common ancestor of modern life are primarily anaerobic, sulfur-reducing hyper thermophiles (White et al. 1995). However, geochemical evidence indicates that Fe (III) is more likely than sulfur to have been the first external electron acceptor of global significance in microbial metabolism (Rich and McKenzie 1981). Archaea and bacteria that are most closely related to the last common ancestor can reduce Fe (III) to Fe (II) and conserve energy to support growth.

© Springer International Publishing Switzerland 2015
S. Mandal, J. Rath, *Extremophilic Cyanobacteria For Novel Drug Development*,
SpringerBriefs in Pharmaceutical Science & Drug Development,
DOI 10.1007/978-3-319-12009-6_1

The most widespread source of reducing power in late Archean and early Protero-zoic (2.9–1.6 Ga) seawater was ferrous iron (0.1–1.0 mM) (Walker 1983), and several authors (Pierson and Olson 1989) proposed that ferrous iron might have been an early electron donor to PS II. The common ancestor of proteobacteria and cyanobacteria might well have used $Fe(OH)^+$ as the principal electron donor for CO_2 fixation (Heis-ing and Schink 1998). Although there is still considerable controversy about the exact time the cyanobacteria started to appear on Earth, there is no doubt that they are ex-tremely ancient organisms. Before oxygen evolution by water splitting, cyanobacteria might well have used $Fe(OH)^+$ as the principal electron donor for CO_2 fixation than other anoxygenic phototrophs. However, the photosystem I (PS I), photosystem II (PS II), and the oxygen evolving complex (OEC) in cyanobacteria appears to have been driven by the necessity to replace $Fe (OH)^+$ with water as an electron donor to fix CO_2 most efficiently in the absence of ferrous iron and leads to evolution of oxygen. Olson 2001 proposed that the driving force for the evolution of PS-II in addition to PS-I was the necessity to utilize $Fe(OH)^+$ effectively for CO_2 fixation in the absence of reduced sulfur compounds. This entailed the evolution of a water-splitting enzyme and the addition of a second photosystem, acting in series with the first, to bridge the enor-mous gap in redox potential between H_2O and NADPH. The present-day structural homologies between photosystems suggest that this change involved the cooperation of a photosystem derived from green bacteria (photosystem I) with a photosystem derived from purple bacteria (photosystem II). The biological consequences of this evolutionary step were far-reaching. For the first time, these cyanobacteria were able to survive with only very minimal chemical demands of their environment. These cells could spread and evolve in ways denied to the earlier photosynthetic bacteria, which needed H_2S or organic acids as a source of electrons. Consequently, large amounts of biologically synthesized, reduced organic materials accumulated. The increase in atmospheric O_2 was very slow at first and would have allowed a gradual evolution of protective devices. The early seas contained large amounts of iron in its ferrous oxida-tion state (Fe^{2+}), and nearly all the O_2 produced by early photosynthetic cyanobacteria was utilized in converting Fe^{2+} to Fe^{3+}. This conversion caused the precipitation of huge amounts of ferric oxides, and the extensive banded iron formations in sedimen-tary rocks beginning about 2.7×10^9 years ago. By about 2×10^9 years ago, the supply of ferrous iron was exhausted, and the deposition of further iron precipitates ceased and moreover, oxygen entered the atmosphere for the first time. Geological evidence suggests that O_2 levels in the atmosphere then began to rise, reaching current levels be-tween 0.5 and 1.5×10^9 years ago. Therefore, this phenomena of evolution of oxygen remarkable shift in biology that required transformations in biochemical mechanisms and metabolic pathways permitted the evolution of new life-forms in which cyano-bacteria play a major role. Much of the internal structure and biochemistry of plastids of higher plants, for instance the presence of thylakoids and particular chlorophylls, is very similar to that of cyanobacteria. According to the endosymbiotic theory, the chlo-roplasts found in plants and eukaryotic algae evolved from cyanobacterial ancestors via endosymbiosis and the phylogenetic estimates constructed with bacteria, plastids, and eukaryotic genomes also suggest that plastids are most closely related to cyano-bacteria. So, cyanobacteria are the key to the evolution of life in this earth.

1.2 Extreme Environments Inhabited by Cyanobacteria and Their Diversity

Cyanobacteria are one of the most successful groups of organisms this planet has ever seen, and include some of the first life forms to evolve on Earth. During their long evolutionary history, cyanobacteria have undergone several structural and functional modifications, and these are responsible for their versatile physiology and wide ecological tolerance. They had to adapt various environmental conditions, from almost anoxic to present oxygen-rich atmosphere, to varying average temperature of the planetary surface, to changes in intensity and spectral composition of light resulting from the changes in composition of the atmosphere and to changes associated with movement to other biotopes. During this process cyanobacteria come up as one of the extremophilic organism. Cyanobacteria occupy almost all the environments on the earth that are illuminated with visible light. Among these habitats there are several places, which are from the anthropocentric view, inhospitable and different from the "normal" places. Extremophiles are classified into various categories according to the "extreme" character of their environments, such as very high or very low temperature limit, pH values, salinity, dryness, high concentration of heavy metals, very high or low levels of radiation, especially ultraviolet radiation, and to a certain extent anaerobic environments. Cyanobacteria are very well equipped to live under extreme environmental conditions such as hyper saline, alkaline, high or low temperature, desiccation, highlight, UV, etc. Contrary to the general belief, many cyanobacteria are not generalists. Many species appear to be highly specialized organisms and can adapt themselves to a narrow set of environmental conditions, particularly in the case of extreme environments. However, our knowledge about the nature of their adaptations is fragmentary. The success of cyanobacteria is not exclusively found in their metabolic diversity, flexibility, and reactivity and they can be diverse and evolve rapidly when conditions change. The diversity of cyanobacteria in various extreme habitats is summarized as follows.

1.2.1 Thermophilic Cyanobacteria

Many geothermal springs emit water near the boiling point. The gradual cooling of the water in the outflow channels provides a stable temperature gradient in which many cyanobacteria positions themselves according to their temperature requirements. It is established that 73–74 °C is the maximum temperature enabling development of cyanobacteria. The existence of thermophilic cyanobacteria has been extensively documented in the course of the microbiological characterization of hot springs, first in Yellowstone National Park, Wyoming, USA (Brock 1969, 1978) and later in other geothermal areas all over the world (Castenholz 1969; Sompong et al. 2005). Different types of unicellular cyanobacteria, classified in the genus *Synechococcus* (*Thermosynechococcus*), are the most thermophilic. There are distinct differences in the communities found in hot springs in different parts of the world. Thermophilic *Synechococcus* species, abundant in Yellowstone, are absent in similar hot springs in Iceland (Ward and Castenholz 2000). Filamentous cyano-bacteria are less thermo-tolerant *Mastigocladus laminosus, Phormidium* sp., and different thermophilic *Oscillatoria* species have their upper temperature limit for growth between 55 and 62 °C where as *Chlorogloeopsis* have the upper temperature limit 65 °C.

1.2.2 Psychrophilic Cyanobacteria

Information on the life of cyanobacteria near the freezing point of water was col-lected in the Antarctic (Vincent 2000). A wide range of species has been found unicellular as well as filamentous. Extensive areas of the McMurdo ice shelf are covered with mats of *Oscillatoria* sp., accompanied by *Nostoc* sp. (Jungblut et al. 2005). Benthic mats lining at the bottom of ice-bound pools in different areas of Antarctica are composed of *Oscillatoria, Lyngbya, Phormidium,* and *Microcoleus*. Cyanobacteria are found in all freshwater environments of Antarctica. *Phormidium* and *Synechococcus* are found in Lake Vanda. *Phormidium frigidum* develops in lakes of the Dry Valleys, sometimes together with *Lyngbya martensiana* (Vincent 2000). The nitrogen-fixing species *Nostoc commune* is abundant in Antarctic soils. Below rocks and in cracks of Antarctic rocks, *Chroococcidiopsis* can often be also found (Vincent 2000). The Antarctic cyanobacteria are not true psychrophiles as most types grow optimally at higher temperatures (15–35 °C) far above of its natu-ral environment. Their growth rate in the cold polar regions is therefore very low. They survive to a large extent based on their tolerance to desiccation, freezing, adapt to low nutrient levels, often to highlight and UV radiation, and the lack of significant levels of predation.

1.2.3 Halophilic Cyanobacteria

Presence of high concentrations of salts does not preclude the occurrence of oxy-genic photosynthesis, and some cyanobacteria thrive in concentrations up to satura-tion. Cyanobacteria are prominently among the phototrophic biota, found in hyper saline environments such as salt lakes, hyper saline lagoons, and solar saltpans. Many highly salt-requiring and salt-tolerant species, unicellular as well as filamen-tous, have been described from such environments. A general review of the occur-rence and properties of halophilic cyanobacteria was reported by Oren (2000). One of the most widely occurring filamentous species is *Microcoleus chthonoplastes*, a benthic mat-building species, found worldwide up to salinities of 200 g/l and higher (Javor 1989). Another type of filamentous cyanobacterium widely encoun-tered in high-salt environments is the coiled *Halospirulina tapeticola* (Nübel et al. 2000). The most widespread and best-known unicellular halophilic cyanobacterium is *Aphanothece halophytica*. In the Great Salt Lake (Utah, USA) cyanobacteria are a characteristic component of the lake's biota. *A. halophytica* is found up to the highest salinities. In addition, filamentous species such as *Phormidium, Oscillatoria* as well as *Microcoleus, Spirulina*, and *Nodularia*, were found in the shallow sediments of the lake (Post 1977). A varied community of cyanobacteria, unicellular as well as filamentous, was described from the hyper saline Solar Lake (Sinai, Egypt; salinity 80–180 g/l), both in the water column and in the benthic microbial mats (Cohen et al. 1977) . Solar salt pans are also a rich source of halophilic cyanobacteria (Javor 1989). At higher salinities *Phormidium, Spirulina, Aphanothece*, and *Synechococcus* become the dominant species of cyanobacteria (Javor 1989). The cyanobacterial community, within the deposits of gypsum, found in saltpan ponds of intermediate salinity has also been well studied (Sørensen et al. 2004). Halophilic and halotolerant cyanobacteria maintain their intracellular ionic concentrations at relatively low levels, although ions such as K^+ and Cl^- can transiently enter into the cells following increases in medium salinity. Many marine and moderately halophilic species, including the abundant *Microcoleus*, produce the organic solutes. Those cyanobacteria adapted to life at the highest salt concentrations (*A. halophytica, Halospirulina*) produce glycine betaine as their osmotic solute. Additional solutes such as L-glutamate betaine have been reported from them (Reed et al. 1986) .

1.2.4 Acidophilic Cyanobacteria

Cyanobacteria generally grow in environments of neutral and alkaline pH, and are rarely found at low pH. Brock (1973) stated in his survey that benthic and planktonic cyanobacteria were never found below pH 4–5, while eukaryotic algae proliferate even at pH-levels below 3.0. However, more recently Steinberg et al. (1998) demonstrated that acid-tolerant cyanobacteria do exist. Populations of two filamentous cyanobacteria resembling *Oscillatoria/Limnothrix* and *Spirulina* sp. were found in acidic Bavarian lakes, which are having a pH of 2.9. Interestingly, eukaryotic phytoplankton was almost absent in that lake. Moreover, a survey of hundreds of lakes in Sweden and Canada showed that cyanobacteria are always present even in the most acidic lakes, down to a pH of about 3.7. Cyanobacteria such as *Aphanocapsa* sp. and several *Chroococcus* sp. have been found to dominate acidified Canadian lakes (Steinberg et al. 1998).

1.2.5 Alkaliphilic Cyanobacteria

The soda lakes of East Africa provide ample documentation for the existence of cyanobacteria adapted to life in highly alkaline environments. In these lakes, *Spirulina platensis* may reach very high community densities and high primary productivity at pH values of 11 and above (Grant and Tindall 1986) . Other species, such as the heterocystous *Anabaenopsis* (*Cyanospira*) and unicellular types such as *Synechococcus* and *Gloeocapsa*, are also reported (Boussiba et al. 2000). *Gloeothece linaris* and *Microcystis aeruginosa* have their optimum pH near 10, and growth of *Plectonema nostocorum* was reported up to pH 13, the highest pH at which life has been recorded. *Spirullina platensis* is an obligate alkaliphile which grows best at pH 9–10 and still grows at 80% of its maximum growth rate at pH 11.5. It has been established that Na^+ is involved in pH homeostasis through the activity of Na^+/H^+ antiporters (Boussiba et al. 2000).

1.2.6 Heavy Metal Tolerant Cyanobacteria

Cyanobacterial cells can be viewed as a natural ion exchanger because they have many anionic groups on their cell surface (Kratochvil and Volesky 1998) and thus enable them to fix metal ions, mainly by means of an ion-exchange mechanism (Schiewer and Volesky 1996). The response of cyanobacteria to toxic metals have been investigated and some of these taxa have been found to display tolerance to toxic metals (Fiore and Trevors 1994). The mesophillic cyanobacteria *Anabaena flos-aquae* and *Synechococcus cedrorum* were grown in various concentrations of carpet industry effluent. The zinc tolerant strain of *Anacystis nidulans* displayed a zinc uptake comparison to a sensitive wild type. *Nostoc microscopicum*, *Nostoc linckia* and *Synechocystis* sp. forms the bloom and are quite tolerant to the toxic metals, copper, cadmium, lead, and zinc.

Several cyanobacterial strains posses, outside of their outer membrane, additional surface layers, mainly of polysaccharidic in nature and referred to as a sheath, capsule, and slimes. Exopolysaccharide producing cyanobacteria could be used for metal biosorption, since most of the polysaccharide envelopes that surround cyanobacterial cells are anionic (De Philippis et al. 2001). Recent studies reported that the capsulated biomass of two filamentous cyanobacteria *Cyanospira capsulata* and *Nostoc* PCC 7936, have the good metal sorption capacity and survive in industrial effluent having multimetal system (De-Philippis et al. 2003). The cyanobacterial cells surrounded by thick polysaccharidic capsules or slime possess a large number of binding sites for metal ions compared to noncapsulated strains. Most cyanobacterial polysaccharides have abundant uronic acid subunits (De-Philippis and Vincenzini 1998) which, owing to their carboxyl groups (Urrutia 1997) efficiently binds metal ions. Thus, the use of the biomass of RPS-producing cyanobacteria for trapping metal ions seems quite promising.

1.2.7 Radiation-Resistant Cyanobacteria

Like other phototrophic organisms, light is obviously essential for cyanobacteria but at very high intensities an unbalanced absorption and utilization of the energy may occur. For instance, exposure to full sunlight at midday may cause over-excitation of the photosynthetic apparatus and cause damage to it as well as to other cellular components. Triplet chlorophyll molecules and oxygen radicals will cause photoinhibition and lead to oxidative damage. This is especially happening when the light includes UV wavelengths. These high-energy shorter wavelengths are deleterious and lead to oxidative stress, DNA damage, and mutations. Many species of cyanobacteria appear to be highly specialized organisms and can tolerate high solar and UV radiation. There are also few cyanobacteria belonging to this genus *Chroococcidiopsis* are extremely resistant to X-ray irradiation (Billi et al. 2000). The bases of the resistance against X-ray irradiation were the capability of Chroococcidiopsis to very effectively and rapidly repair DNA damage. Cyanobacteria originated early during precambrian era, i.e., before the existence of the present ozone shield, hence, it is presumed that they faced more intense solar UVR as compared to other phylogenetically much younger phototrophs. UV-B exposure had negligible short-term effects on the growth of *Nostoc commune* (Ehling-Schulz et al. 1997). Quesada et al. (1999) found decrease in phycocyanin/chlorophyll-a and increase in carotenoid /chlorophyll -a with response to UV-A and UV-B in *Phormidium murrayi* a mat forming cyanobacterium isolated from ice self pond in Antarctica and even after 6 h of UV radiation there was almost no change in its morphological features. *Lyngbya aestuarii* from Chilika Lake, India is also quite tolerant to UV-B irradiation and have an efficient adaptation strategy to tolerate, long-term UV-B irradiation (Rath and Adhikary 2007). *Lyngbya majuscule* another mat forming cyanobacteria from saline soil also tolerant to UV-B and its metabolic activities were considerably revived after incubating the irradiated cells in mineral medium under fluorescent light and in the dark suggesting existence of photoreactivation and dark repair in this cyanobacterium (Mandal et al. 2011).

Radiation damage can be attributed by reactive oxygen species (such as singlet oxygen) formed at high light intensities. Cyanobacteria often grow on walls and pavements of building and historic monuments are exposed to full sunlight. Some of the most extreme levels of radiation that cyanobacteria encounter are found in Antarctica, where degradation of the ozone layer has brought about an increase in the amount of solar ultraviolet radiation that reaches the surface (Vincent 2000). Chlorophyll-a is the essential molecule for nearly all oxygenic photosynthetic organisms, from cyanobacteria to higher plants, excluding the Chlorophyll d-containing cyanobacterium, *Acaryochloris marina* (Chen et al. 2005). *Acaryochloris marina* is the only cyanobacterium reported that uses chlorophyll-d as its major photosynthetic photopigment. It is found in filtered light environments in various ecological niches. The advantage of using Chlorophyll-d and long wavelength absorbing chlorophylls in cyanobacteria is intriguing due to its unique absorption properties and its potential for increased photosynthetic efficiency. *Acaryochloris marina* uses Chlorophyll-d, up to 95–99 % of total chlorophylls, as its major photopigment (Miyashita et al. 1996). Chlorophyll-d can replace all functions of Chlorophyll-*a* in *A. marina* not only in light-harvesting complexes (Chen et al. 2002; Tomo et al. 2011), but also in reaction centers (Chen et al. 2005; Tomo et al. 2007). Recently chlorophyll gets an *f* from cyanobacteria. The newly discovered cyanobacterium *Halomicronema hongdechloris* which contains Chlorophyll *f* was isolated from stromatolites found in Shark Bay, Western Australia, and cultured in the laboratory (Li et al 2014). Chlorophyll *f* from this cyanobacterium has a maximum Q_Y absorption peak at about 706 nm and a maximum fluorescence emission at 722 nm at room temperature in methanol, making it the most red-shifted chlorophyll discovered to date (Chen et al. 2010). This finding suggests that cyanobacterial photosynthesis can be extended further into the infrared region.

1.3   Survival Strategy of Extremophilic Cyanobacteria

Cyanobacteria have developed an array of exceptional qualities and responses to ensure their survival in the extreme climatic conditions in which they grow. The presence of so many unique characteristics in some of these organisms creates difficulties for researchers to study only one trait without considering the effects of another. Cyanobacteria use mostly three different types of strategies to counteract extreme environmental conditions, these are: (a) stress avoidance mechanisms, (b) stress defense activities, and (c) repair mechanisms including DNA repair (Pattanaik et al. 2007).

1.3.1   Avoidance Mechanism

Under the avoidance mechanism extremophilic cyanobacteria follows four different adaptation strategies to counteract stress, i.e., migration, mat formation, change in morphology, and formation of extracellular polysaccharides. Cyanobacteria are the most successful mat forming organisms. They are closely associated with the substrate-producing mat like structure. These are of various thicknesses ranging from a few micrometers to a few millimeters. The mats are composed of a varying number of different cyanobacteria (10–40 species) (Büdel 1999). They have wide range of metabolic capabilities to adapt to extreme environmental conditions (high temperature, high light intensity, high humidity, and low water availability), prevailing in alkaline hot springs (Miller et al. 1998), Arctic fresh waters (Quesada et al. 1999), hot, arid areas (Scherer et al. 1984), rock surfaces of hot deserts (Friedmann 1967), on the bark surface (Sinha et al. 1999), rice fields (Adhikary and Sahu 2000), and on exposed rock surfaces (Pattanaik and Adhikary 2002). Generally large filamentous cyanobacteria colonize first. They possess thick sheath or slime around the filament. Composition of the microbial mat varies according to different environmental conditions and the nature of the substratum. Under extreme conditions they develop morphological and physiological adaptation that allow them to remain viable and dormant in desiccated state and under high temperature during summer months (Kovacik 2000). According to Belnap et al. (2001), the crust formation is a result of intimate association between the substrate and the microorganisms, which live within or immediately on the top of the uppermost layer. Dry cyanobacterial crusts were moistened osmotically at different water potential (Pattanaik et al. 2007).

To escape from different extreme stress condition, cyanobacteria in the mat often migrate downward and vertically depending upon the conditions of the environment. Ramsing and Prufert–Bebout (1994) reported downward movement of motile oscillatorian cyanobacteria from microbial mat surfaces into the mat matrix or into soft sediments during period of high solar irradiance. Sinking and floating behavior by gas vacuoles in the cyanobacteria act as a protective strategy against UV radiation (Reynolds 1987). In Antarctic cyanobacteria there are substantial differences between closely related species to escape damaging effects. During continuous daylight of the Antarctic summer, the motile trichomes of *Oscillatoria priestleyi* remain at the bellow of mat surface and only rise to the surface if the mat is subjected to prolonged hours of shading (Vincent and Quesada 1994). Daily vertical migration to avoid periods of incident high solar irradiance has been reported for *Oscillatoria* species and *Spirulina* ef. *subsala* (Garcia–Pichel et al. 1994). UV irradiance is the primary cause of the vertical movement of cyanobacteria (Garcia–Pichel and Castenholz 1994). Downward movement occurred not only in UV-B but in response to high intensities of UV-A (more than 10 W/m^2) and broad visible irradiance over (400 W/m^2).

Stress can also affect the motility and photo-orientation of cells and alter the morphology through breaking and changing the cyanobacterial filament to spiral or coiled structure (Rath and Adhikary 2007). However, cyanobacteria living under high levels of solar radiation, implying that they must possess efficient mechanisms to counteract the harmful effects of UVR. Though there are few studies on physiological responses of cyanobacteria to UV radiation, little has been documented on morphological modifications caused by UVR (Mandal et al. 2011). Significantly shorter trichomes were found in cultures exposed to high solar radiation. However, most of the breakage and spiral alteration occur with UV-B (Gao et al. 2007). The damage caused by UVR decreases with increase in temperature ranging from 15 to 30 °C (Gao et al. 2008). However, Zengling and Gao 2009 demonstrated that it was PAR which caused the spiraling of *A. platensis* to tighten and the presence of UVR accelerated this compressing process suggested that the self-shading provided by the compress spirals to photoprotect against harmful radiation (Wu et al. 2005).

Several cyanobacterial strains possess, outside their outer membrane, additional surface layers, mainly of polysaccharide in nature and referred to as sheath, capsule, and slimes. During cell growth in batch cultures, aliquots of the polysaccharidic material of both capsule and slimes may be released as water-soluble material into the surrounding medium, causing a progressive increase of its viscosity. These water soluble released polysaccharides (RPSs), being easily recovered from liquid culture are currently attracting much interest because of their suitability for a variety of industrial purposes. Cyanobacteria have batter survival capacity in desiccated condition than the other microorganisms as they release extracellular polysaccharides (EPS) when exposed to high light and desiccation which produce a matrix that stabilizes the sediment (Helm et al. 2000). The quick absorption and slow loss of water at high and low water availability have been attributed to gel like protoplasm which consequently serve as a protective mechanism to overcome long periods of drought (Potts and Friedmann 1981). The EPS—containing sheath of cyanobacteria forms a buffer zone between the environment and the cell. There were reports that fibrilar sheath and diffuse slime layer found in many cyanobacteria are mostly composed of carbohydrates and provide protection under stress condition (Adhikary et al. 1986; Weckesser et al. 1987).

1.3.2   Stress Defense Activities

In response to various stresses cyanobacteria change their metabolic activity and synthesize various chemicals, which often summarize as chemical defense strategies. There are various chemical defense strategies of cyanobacteria and are antioxidant defense system, synthesis of stress proteins, synthesis of photo protective compounds, and production of an array of secondary metabolites. Cyanobacteria have evolved a complex defense system against reactive oxygen species (ROS) including nonenzymatic antioxidants like carotenoids, tocopherols (vitamin E), ascorbic acid (vitamin C) and reduced glutathione. Enzymatic antioxidants are superoxide dismutase (SOD), catalase and glutathione peroxidase as well as the enzymes involved in the ascorbate glutathione cycle, such as ascorbate peroxidase (APX), mono-dehydroascorbate reductase, dehydroascorbate reductase, and glutathione reductase. Carotenoids are well known for their antioxidant activity (Tarko et al. 2012). In cyanobacteria, carotenoids occur in the outer cell membrane as well as in thylakoids. During long-term exposure to high natural or artificial radiation the carotenoids and chlorophyll-a ratio become very high. This high ratio is the prerequisite for a broader tolerance against excessive irradiation, particularly at suboptimal temperature (Castenholz 1972), and in response to UVR (Ehling–Schulz and Scherer 1999). To counteract UV-B induced ROS, an increase in the synthesis of carotenoids in *Microcystis aeruginosa* was observed (Jiang and Qiu 2005). Carotenoids exerted their protective function as antioxidants in *Synechococcus* PCC7942 by inactivating free radicals in the photosynthetic membrane (Goetz et al. 1999). It is also reported that outer membrane bound carotenoids provided a fast active SOS response to counteract acute cell damage (Ehling- Schulz et al. 1997). The nonenzymatic antioxidants are not considered as the efficient detoxifying agents (Wolfe- Simon et al. 2005). The most important enzymes, which detoxify superoxide radicals, are the SOD family, eliminating the noxious superoxide radical anions. Different metalloforms of SOD (Fe, Mn, Cu, Zn, and Ni) protect different cellular proteins and provide an *in vivo* tool for studying cellular responses to oxidative stress (Lesser and Stochaj 1990). Under desiccation, rehydration and during UVA and UVB irradiation, an accumulation of active Fe-SOD was detected in *Nostoc commune* (Ehling- Schulz et al. 2002). APX effectively removes low concentrations of peroxides whereas catalase eliminates H_2O_2 produced under photo-oxidative conditions. High activities of APX and catalase were reported in *Nostoc muscorum* 7119 and *Synechococcus* sp. 6311 (Miyake et al. 1991). One and two-years-old dry mats and the corresponding organism *Lyngbya arboricola* exhibited an enhancement in the activity of active oxygen scavenging enzymes superoxide dismutase (SOD), APX and catalase in their cell (Tripathy and Srivastava 2001).

Cyanobacterial species accumulate a small set of proteins whenever exposed to any stress. The desiccation tolerant cyanobacterium *N. commune* is one such organism that accumulates novel group of water stress proteins (Wsp; Hill et al. 1994). These proteins occur in high concentration showing isoelectric points between 4.3 and 4.6 with molecular masses of 33, 37, and 39 kDa. Wsp polypeptides and UV-A/B absorbing compounds secreted by cells accumulate in the extracellular glycan sheath and are released from desiccated colonies upon rehydration (Hill et al. 1994). According to Scherer and Potts (1989), the amino terminal sequence of the 39 kDa protein is Ala-Leu-Tyr-Gly-Tyr-Thr-Ile-Gly-Glu. Peptide mapping of the 39 and 33 kDa proteins using different proteases gave a similar pattern of digestion fragments suggesting that this protein is a water stress protein with a protective function on the structural level. The knowledge about the effects of UV at the protein level is limited. An induction of UV-shock proteins in response to high intensities of UV (295–390 nm) and in response to UV-C (265 nm) irradiation has been reported in cyanobacteria (Shibata et al. 1991). Heat shock proteins (HSPs) are synthesized if the organisms are exposed to temperatures above the optimum level (Neidhart et al. 1984). The structure of HSPs is highly conserved and their function is similar in all organisms. HSPs serve in the cell as molecular chaperones, i.e., facilitate correct protein folding and degradation of unfolded or denatured proteins. If the cells are exposed to the heat shock, the requirement for HSPs increases for folding of the newly synthesized polypeptides and degradation of the heat denatured proteins. The HSPs synthesis is also induced by other stresses, e.g., by the presence of toxic compounds, but the sudden exposure of the cyanobacteria to low temperatures does not induce their synthesis. During temperature shift to low temperatures, the reactions of the cold shock occur and cold shock proteins (CSPs) are synthesized during the initial lag-phase.

Cyanobacteria are known to produce photo-protecting compounds to protect themselves from high solar and UV radiation. Photoprotective compounds like mycosporine-like amino acids (MAAs) and scytonemin play an important role in allowing these organisms to survive in habitats exposed to strong irradiation. MAAs are colorless, water-soluble, low molecular weight compounds, which accumulate in large quantities in cyanobacterial cells (Karsten and Garcia- Pichel 1996). Structurally, these are cyclohexenone or cyclohexenimine chromophores conjugated with the nitrogen substituent of an amino acid or its imino alcohol, having absorption maxima ranging from 310 to 360 nm and an average molecular weight of around 300 Da (Cockell and Knowland 1999). Their biosynthesis probably originates from the first part of the shikimate pathway via 3-dehydroquinic acids and 4-deoxygadusol (4-DG; Favre- Bonvin et al. 1987). Till today about 20 MAAs have been reported from various cyanobacterial species growing in different habitats. The most common MAAs are shinorine, asterina-330, porphyra-334, and mycosporine-glycine. Several new compounds have also been characterized in recent years such as the novel compound 2-(E)-3-(E)-2, 3-dihydroxyprop-1-enylimino-mycosporinealanine in the unicellular cyanobacterium *Euhalothece* sp. strain LK-1, isolated from a gypsum crust of a hypersaline saltern pond in Eilat, Israel (Volkmann et al. 2006).

MAAs serves multiple roles in the cellular metabolism in cyanobacteria. According to Sinha et al. 1999, MAAs may serve at least three different functions in the cyanobacteria, (i) protection against deleterious UV radiation as a sunscreen, (ii) transfer radiant energy to the photosynthetic reaction centers, which is supported by the emitted fluorescence spectrum peaks at a wavelength near the soret band of chlorophyll absorption, and (iii) aid in osmotic regulation. The protection against UV damage by MAAs depends on the species and the location of the pigment. In *N. commune* MAAs are thought to play an important role in photo protection because MAAs are located in the extracellular glycan covalently linked to oligosaccharides (Ehling- Schulz et al. 1997).

The other UV absorbing compound is known as scytonemin, which is also known for its photo protective role. It is an extracellular, yellowish-brown, lipid-soluble dimeric pigment with a molecular weight of 544 Da and a structure based on indolic and phenolic subunits. Thick yellowish-brown sheath layer or slime layer mostly surrounds cyanobacteria occurring in various harsh environmental conditions including high solar radiation and desiccation. Nägeli (1849) first described the brown coloration of cyanobacteria, in microbial mats, due to the presence of scytonemin. Scytonemin may have evolved during the Precambrian era and colonize at exposed terrestrial habitats in cyanobacteria or their oxygenic ancestor (Dillon and Castenholz 1999). Scytonemin is formed by condensation of tryptophan and phenyl-propanoid derived subunits (Proteau et al. 1993). The biochemical pathways that are involved in the biosynthesis are still unknown. However, scytonemin synthesis is strongly induced by UV-A and due to its high extinction coefficient it may prevent 90 % of incident UV-A entering the cell, thus, efficiently functioning as an UVA-sunscreen (Brenowitz and Castenholz 1997).

1.3.3 Repair Mechanism

Repair is another important mechanism in cyanobacteria to persist in nature under different stress (Castenholz 1997). The repair mechanism involves active processes, in which the damaged molecules are replaced by synthesis of the target molecules or by repair of damaged targets without de novo synthesis (i.e., in DNA repair; Pattanaik et al. 2007). During photoreactivation cyclobutane type pyrimidine dimers are monomerized by the enzyme DNA photolyase which is activated by UV-A and blue light (Pang and Hays 1991). Excision repair is light-independent and various enzymes are involved. Cyanobacteria have been found to exhibit both photo reactivation and excision repair (Eker et al. 1990). *RecA*-like genes from cyanobacteria have been shown to complement a *recA* deletion in *Escherichia coli* (Geoghegan and Houghton 1987) and show its function by efficiently repair cellular functions occurred by various stresses.

1.4 Changes in Metabolic Activities of Cyanobacteria Under Stress

Environmental stresses influence the physiological activities of many cyanobacteria. Cyanobacterial cell under stress, (changes in temperature, salinity, pH, etc.) first sensed by the specialized sensory proteins, or sensors, which change their properties under the stress. Sensors transfer the signal about the changes to other polypeptides—the transducers, which in turn regulate expression of the stress responsive genes. Transducer proteins may recognize the special regions of DNA directly, interact with them, and, in such way, regulate transcription. Finally, protective proteins and/or metabolites are synthesized that help the cells to adapt or acclimate to the new environment. In response to moderate stress, many cyanobacteria activate sets of genes that are specific to the individual type of stress. The specific proteins that are synthesized and some of these proteins, in turn, participate in the synthesis of certain stress-specific metabolites. The cosmopolitan distribution of cyanobacteria indicates that they can cope with a wide spectrum of global environmental stresses such as heat, cold, desiccation, salinity, nitrogen starvation, photo-oxidation, an aerobiosis and osmotic stress, etc. (Sinha and Häder 1996). They have developed a number of mechanisms by which defend themselves against environmental stresses. Important among them are the production of photoprotective compounds such as MAAs and scytonemin (Sinha et al. 2001), enzymes such as superoxide dismutase, catalases and peroxidases (Canini et al. 2001) and synthesis of stress proteins (Sinha and Häder 1996). However, the metabolic activities of each genus and even strain are different in different stress conditions. Some desiccation-tolerant cyanobacteria accumulate large amounts of disaccharides trehalose and sucrose (Crowe and Crowe 1992) and are effective at protecting enzymes during both freeze-drying and air drying. Molecular modeling shows that trehalose can fit between the phosphate of adjacent phospholipids (Rudolph et al. 1990). In low trehalose/lipid ratios trehalose is not available to bind water, thus showing a direct interaction between the sugar and lipid. Cyanobacteria also have several kinds of metabolic activities that allow them to acclimate to salt stress. The inducible synthesis of compatible solutes such as sucrose is synthesized in salt-sensitive strains of cyanobacteria such as *Synechococcus* (Hagemann and Erdmann 1997); glucosylglycerol is synthesized in strains with intermediary tolerance such as *Synechocystis* sp. PCC 6803 (Hagemann et al. 2001); glycinebetaine is synthesized in salt tolerant *Synechococcus* sp. PCC 7418 (Hagemann and Erdmann 1997). Many reports have suggested that lipids might be involved in the protection against salt stress (Ritter and Yopp 1993). Cyanobacteria respond to a decrease in ambient temperature by desaturating the fatty acids of membrane lipids to compensate for the decrease in membrane fluidity at low temperatures (Murata and Nishida 1987). Fatty acid desaturases are the enzymes that introduce the double bonds into the hydrocarbon chains of fatty acids, and thus these enzymes play an important role during the process of cold acclimation of cyanobacteria (Wada and Murata 1990). Cyanobacteria on, exposed to high photo synthetically active radiation (PAR) or UV radiation leads to photoinhibition of photosynthesis by limiting the efficient fixation of light energy (Nishiyama et al. 2001). The molecular mechanism of photo inhibition revealed that the light-induced damage is caused by inactivation of the D1 protein of the PSII complex (Tyystjärvi et al. 2001). Therefore, the ability of membrane lipids to desaturate fatty acids is important for them to tolerate high light stress, by accelerating the *de novo* synthesis of the D1 protein. Because of the tolerance to various stressed environments the protective proteins and/or metabolites are synthesized in them and they emerge as an ideal candidate for drug discovery. A majority of these metabolites are peptides and are synthesized by large multi modular nonribosomal polypeptide (NRPS) or mixed polyketide (PKS)-NRPS enzymatic systems (Schwarzer et al. 2003). Thus, cyanobacteria continue to be explored and their metabolites are now evaluated in a number of biological areas and they are becoming an exceptional source of leading compounds for drug discovery (Singh et al. 2005; Nunnery et al. 2010).

1.5   Cyanobacteria as a Promising Candidate for New Pharmaceutical Compounds

In the microbial world, cyanobacteria are prolific producers of secondary metabolites, many of which show various biological activities or bioactivity. Gerwick et al. (2008) found that secondary metabolites are mostly isolated from the members of oscillatoriales (49%), followed by nostocales (26%), chroococcales (16%), pleurocapsales (6%) and stigonematales (4%). Cyanobacteria such as *Anabaena, Nostoc, Microcystis, Lyngbya, Oscillatoria, Phormidium* and *Spirulina* produce a variety of high value compounds such as carotenoids, fatty acids, lipopeptides, polysaccharides, and other bioactive compounds. Cyanobacterial secondary metabolites include different compounds like cytotoxic (41%), antitumor (13%), antiviral (4%), antimicrobial (12%) and other compounds (18%) include antimalarial, antimycotics, multidrug resistance reversers, antifeedant, herbicides, and immune suppressive agents (Burja et al. 2001).

The tropical marine environments and freshwater habitats are the two most widely studied habitats for cyanobacterial metabolites. Notably, compounds isolated from these different environments show distinct metabolite profiles, with respect to both structural properties and bioactivity features. Strains from tropical marine habitats such as *Lyngbya majuscula* frequently produce cytotoxic compounds, some of them having high potential for drug development. The structures of these metabolites, such as curacin A, jamaicamide, barbamide, hectochlorin, lyngbyatoxin, and aplysiatoxin, possess a polyketide backbone, but often contain amino acid constituents (Jones et al. 2009). On the other hand, bloom forming freshwater cyanobacteria such as *Microcystis, Planktothrix*, and *Anabaena* produce a variety of peptide backbones, eventually having polyketide side chains. Examples include microcystin, aeruginosin, anabaenopeptin, cyanopeptolin, microginin, and microviridin showing activity against various proteases or protein phosphatases (Welker and Dohren 2006). Cyanobacterial strains from terrestrial sources were only incidentally screened, but revealed different types of compounds, such as cryptophycin, nostopeptolide and nostocyclopeptide, scytonemin, and mycosporic like amino acid (Kehr et al. 2011). Although certain types of cyanobacterial compounds, including microcystin and nostopeptolide, have been detected in different habitats the majority of them seem to be specific to a particular environmental niche, pointing toward distinct ecological and physiological roles for these compounds.

Cyanobacteria are able to survive in metabolically extreme environments, where the energy flow is low, by exploiting both the electron donor and electron acceptor, which are not available to other eukaryotes, and might be the causes of extraordinary capacities to produce structurally diverse and highly bioactive secondary metabolites. Because of the physical and chemical conditions in the extreme environment, extremophilic cyanobacteria exhibit a variety of molecules with unique structural features. Genomic studies show, they are having multiple biosynthetic pathways encoding bioactive secondary metabolites, suggesting that natural product richness is an evolutionary adaptation in them (Engene et al. 2013). To date, an impressive 600 natural products (NPs) have been reported from marine strains of cyanobacteria. The tropical and subtropical, benthic marine cyanobacteria are considered a promising source for new pharmaceutical lead compounds and characteristic of high chemical diversity, biochemical specificity and high binding affinities to their specific receptor and having significant pharmacological activities specially in the field of cancer and infectious disease. The majority of bioactive metabolites isolated from cyanobacteria have either been polyketides, nonribosomal peptides, or a hybrid of the two. Nonribosomal peptide synthetase genes are very ancient part of the cyanobacterial genome and presumably have evolved by recombination and duplication events to reach the present structural diversity of cyanobacterial oligopeptides. The high diversity of cyanobacterial secondary metabolites and their chemical structures indicates the presence of diverse NRPS and PKS gene clusters in cyanobacterial genomes, though only a minor part has been sequenced so far.

An advantage of natural products research on marine cyanobacteria is the high discovery rate (>95 %) of novel compounds as compared to other traditional microbial sources. One of the key areas to further tap these cyanobacteria for new chemical entities is the collection of cyanobacterial strains from unexplored localities, especially from Africa and Asia. Systematic research during the last 4 years has led to the isolation and identification of several new bioactive compounds of potential therapeutic use. Most of the anticancer peptides have been reported from predominantly two genera: *Lyngbya* and *Leptolyngbya* sp. Bisebromoamide is a novel terminal-protected, linear heptapeptide consisting of D-amino acids, N-methyl amino acids and modified amino acids, obtained from the Okinawan strain of *Lyngbya* sp. This compound was active against a panel of 39 human cancer cell lines and HeLa S3 cells with a mean IC_{50} of 40 nM (Teruya et al. 2009). It was found that bisebromoamide selectively inhibits the phosphorylation of ERK (extracellular signal regulated protein kinase). Aberrant activation of Ras/Raf/MEK/ERK pathway is commonly observed in various cancers. Small molecules targeting these pathways are much sought after cancer treatment and bisebromoamide could be a lead molecule for this type of therapy in the future. Recently, synthesis of bisebromoamide has also been accomplished. A Fijian collection of *Lyngbya majuscula* yielded majusculamide C, a cyclic peptide of both L- and D-amino acids, partial N-methylations, and an unusual γ-amino acid, together with its open chain analog. Both the cyclic as well as acyclic metabolites were active against HCT-116 cell line at very low nanomolar concentrations (IC_{50}=20 and 16 nM, respectively; Simmons et al. 2009). Even at low dosage of 52 nM, these compounds caused complete disruption of actin filaments and induced dramatic changes to the cell morphology. *Lyngbya majuscula* from Singapore afforded the cyclodepsipeptides, hantupeptins A–C, of hybrid polyketide peptide origin (Tripathi et al. 2010). Among these, hantupeptin A was active against MOLT 4 leukemia cell lines at nano molar levels (IC_{50}=32 nM). Lagunamides A and B, new cyclodepsipeptides from *L. majuscula*, collected from Pulau Hantu Besar, Singapore, were active against P388 murine leukemia cell lines (IC_{50}=6.4 and 20.5 nM, respectively). Lagunamide A has comparable antimalarial activity as dolastatin-15, though it is lower than dolastatin- 10, the most potent antimalarial agent of cyanobacterial origin known today. Dragonamide E, a linear peptide of *L. majuscula* from Panama exhibited antileishmanial activity (Balunas et al. 2010). A lariat-type cyclic depsipeptide, coibamide A, active at nanomolar levels against the majority of the NCI's panel of 60 cancer cell lines was reported from a *Leptolyngbya* strain from Panama (Medina et al. 2008). It is proposed that perhaps this compound is acting in a novel way and efforts are on to synthesize the molecule in the lab. Another macrocyclic compound of potential biomedical use is palmyrolide A, isolated from a mixed cyanobacterial assemblage of *Leptolyngbya* cf. and *Oscillatoria* sp. This macrocyclic lipophilic compound effectively blocks sodium channel (Pereira et al. 2010).

Antiviral compound isolated from cyanobacteria are usually found to show bioactivity by blocking viral absorption or penetration and inhibiting replication stages of progeny viruses after penetration into cells. A new class of HIV inhibitors called sulfonic acid, containing glycolipid, was isolated from the extract of cyanobacteria and the compounds were found to be active against the HIV virus. Cyanovirin-N (CVN), a peptide isolated from cyanobacteria, inactivates the strains of HIV virus and inhibits cell to cell and virus to cell fusion (Yang et al. 1997). *In vitro* and *in vivo* antiviral tests suggested that the anti-HIV effect of CVN is stronger than a well-known targeted (viral entry) antibody (2G12) and another microbicide, PRO2000 (Xiong et al. 2010). Yakoot and Salem (2012) have conducted first human trial to address the effect of *Spirulina platensis* dried extract on virus load, liver function, health related quality of life and sexual functions in patients with chronic hepatitis C virus (HCV) infection. They found the therapeutic potential of *S. platensis* in chronic HCV patients, and in some cases (13%) the viral infection is complexly nullified. Mansour et al. (2011) have found that the polysaccharides isolated from *Gloeocapsa turgidus* and *Synechococcus cedrorum* showed higher antiviral activity against rabies virus than that against herpes-1 virus. The exopolysaccharide from *Aphanothece halophytica* has an antiviral activity against influenza virus A (H1N1), which shows a 30% inhibition of pneumonia in infected mice (Zheng et al. 2006).

The drug-development process normally proceeds through various phases of clinical trials (phase 0 or pre-clinical, phase I, phase II and phase III). The FDA must approve each phase before the study can continue. There are few prominent molecules from cyanobacteria such as dolastatins, cryptophycins, curacin, and their analogues, which are in clinical trials as potential anticancer drugs, and many novel molecules are in the process of clinical trials. Many promising natural products could not be developed into potential therapeutic agents owing to supply problems. Efforts are now on to identify the "real" microbial source of several promising drug molecules and further, to develop suitable fermentation strategies or chemical synthesis to ensure their continued supply in future. In this context, cyanobacteria are found to be a promising source for novel therapeutic agents.

References

Adhikary SP, Sahu JK (2000) Survival strategies of cyanobacteria occurring as crust in the rice fields under drought conditions. Ind J Microbiol 40:53–56

Adhikary SP, Weckesser J, Jürgens UJ, Golecki JR, Borowia D (1986) Isolation and chemical characterization on the sheath from the cyanobacterium *Chroococcus minutus* SAG B. 41.79. J Gen Microbiol 132:2595–2599

Balunas MJ, Linington RG, Tidgewell K, Fenner AM, Urena LD, Della Togna G et al (2010) Dragonamide E, a modified linear lipopeptide from lyngbya majuscula with antileishmanial activity. J Nat Prod 73:60–66

Belnap J, Prasse R, Harper KT (2001) Influence of biological soil crust on soil environments and vascular plants. In: Benlap J, Lange OL (eds.) Biological soil crusts: structure, function, and management. Springer, Berlin, pp 281–300

Billi D, Friedman EI, Hofer KG, Grilli Caiola M, Ocampo- Friedmann R (2000) Ionization- radiation resistance in the desiccation- tolerant cyanobacterium Chroococcidiopsis. Appl Environ Microbiol 66:1489–1492

Boussiba S, Wu X, Zarka A (2000) Alkaliphilic cyanobacteria. In: Seckbach J (ed) Journey to diverse microbial worlds: adaptation to exotic environments. Kluwer Academic, Dordrecht, pp 209–224

Braterman PS, Cairns-Smith AG, Sloper RW (1983) Photo-oxidation of hydrated Fe^{2+} significance for banded iron formations. Nature 303:163–163

Brenowitz S, Castenholz RW (1997) Long- term effects of UV and visible irradiance on natural populations of scytonemin containing cyanobacterium (*Calothrix* sp.). FEMS Microbiol Ecol 24:343–352

Brock TD (1969) Microbial growth under extreme conditions. Symp Soc Gen Microbiol 19:15–41

Brock TD (1973) Lower pH limit for the existence of blue- green algae: evolutionary and ecological implications. Science 179(72):480–483

Brock TD (1978) Thermophilic Microorganisms and life at high temperature. Springer verlag, New York, pp 465

Büdel B (1999) Ecology and diversity of rock inhabiting cyanobacteria in tropical regions. Eur J Phycol 34:361–370

Burja AM, Banaigs EB, Abou-Mansour, Burgess JG, Wright PC (2001) Marine cyanobacteria-a prolific source of natural products. Tetrahedron 57:9347–9377

Canini A, Leonardi D, Caiola MG (2001) Superoxide dismutase activity in the cyanobacterium *Microcystis aeruginosa* after surface bloom formation. New Phytol 152:107–116

Castenholz RW (1969) Thermophilic blue- green algae and the thermal environment. Bacteriol Rev 33:476–504

Castenholz RW (1972) Low temperature acclimation and survival in thermophilic *Oscillatoria terebriformis*. In: Desikachary T.V. (ed.) Taxonomy and biology of blue- green algae. University of Madras, India, pp 406–418

Castenholz RW (1997) Multiple strategies for UV tolerance in cyanobacteria. Spectrum 10:10–16

Chen M, Quinnell RG, Larkum AWD (2002) The major light-harvesting pigment protein of *Acaryochloris marina*. FEBS Lett 514:149–152

Chen M, Telfer A, Lin S, Pascal A, Larkum AWD, Barber J, Blankenship RE (2005) The nature of the photosystem II reaction centre in the chlorophyll d-containing prokaryote, *Acaryochloris marina*. Photochem Photobiol Sci 4:1060–1064

Chen M, Schliep M, Willows RD, Cai ZL, Neilan BA, Scheer H (2010). A red-shifted chlorophyll. Science 329(5997):1318–1319

Cockell CS, Knowland J (1999) Ultraviolet radiation screening compounds. Biol Rev 74:311–45

Cohen Y, Krumbein WE, Goldberg M, Shilo M (1977) Solar lake (Sinai). 1. Physical and chemical limnology. Limnol Oceanogr 22:597–608

Crowe JH, Crowe LM (1992) Anhydrobiosis: a strategy for survival. Adv Space Res 12:239–247

De Philippis R Vincenzini M (1998) Exocellular polysaccharides from cyanobacteria and their possible applications. FEMS Microbiol Rev 22:151–175

De Philippis R Sili C Paperi R Vincenzini M (2001) Exopolysaccharide- producing cyanobacteria and their possible exploitation: a review. J Appl Phycol 13:293–299

De Philippis R Paperi R Sili C Vincenzini M (2003) Assessment of the metal removal capability of two capsulated cyanobacteria *Cyanospira capsulata* and *Nostoc* PCC7936. J Appl Phycol 13:293–299

Dillon JG, Castenholz RW (1999) Scytonemin, a cyanobacterial sheath pigment, protects against UVC radiation: implications for early photosynthetic life. J Phycol 35:673–681

Ehling- Schulz M, Scherer S (1999) UV protection in cyanobacteria. Eur J Phycol 34:329–338

Ehling- schulz M, Bilger W, Scherer S (1997) UV-B induced synthesis of photoprotective pigments and extra cellular polysaccharides in the terrestrial cyanobacteria *Nostoc commune*. J Bacteriol 179:1940–1945

Ehling- Schulz M, Schulz S, Wait R, Görg A, Scherer S (2002) The UV-B stimulon of the ter-restrial cyanobacterium *Nostoc commune* comprises early shock proteins and late acclimation proteins. Mol Microbiol 46:827–843

Eker APM, Kooiman P, Hessels JKC, Yasui A (1990) DNA photoreactivating enzyme from the cyanobacterium *Anacystis nidulans*. J Biol Chem 265:8009–8015

Engene N, Gunasekera S, Gerwick WH, Paul VJ (2013) Phylogenetic inferences reveal large extent of novel biodiversity in chemically rich tropical marine cyanobacteria. Appl Environ Microbio 79:1882–1888

Favre- Bonvin J, Bernillon J, Salin N, Arpin N (1987) Biosynthesis of mycosporines: mycosporine glutaminol in *Trichothecium roseum*. Phytochemistry 26:2509–2514

Fiore MF, Trevors JT (1994) Cell composition and metal tolerance in cyanobacteria. Biometals 7:83–81

Friedmann EI, Lipkin Y, Ocampo- Paus R (1967) Desert algae of the Negev (Israel). Phycologia 6:185–200

Gao K, Yu H, Brown MT (2007) Solar PAR and UV radiation affect the physiology and morphology of the cyanobacterium *Anabaena* sp. PCC 7120. J Photochem Photobiol B Biol 89:117–124

Gao KS, Li P, Walanabe T, Helbling EW (2008) Combined effects of ultraviolet radiation and temperature on morphology, photosynthesis, and DNA of *Arthrospira* (*Spirulina*) platensis (*Cyanophyta*). J Phycol 44:777–786

Garcia- Pichel F, Castenholz RW (1994) On the significance of solar ultraviolet radiation for the ecology of microbial mats. In: Stal LJ, Caumette P (eds) Microbial mats, structure, develop-ment and environmental significance, NATO ASI series. Springer, Berlin, pp 77–84

Garcia- Pichel F, Mechling M, Castenholz RW (1994) Dial migrations of microorganisms within a benthic, hypersaline mat community. Appl Environ Microbiol 60:1500–1511

Geoghegan CM, Houghton JA (1987) Molecular cloning and isolation of a cyanobacterial gene which increases the UV and methyl methanesulphonate survival of recA strains of *Escherichia coli* K12. J Gen Microbiol 133:119–126

Gerwick WH, Coates RC, Engene N, Gerwick LG, Grindberg R, Jones A, Sorrels C (2008) Giant marine cyanobacteria produce exciting potential pharmaceuticals. Microbe 3:277–284

Goetz T, Windhoevel U, Boeger P, Sandmann G (1999) Protection of photosynthesis against ultraviolet- B radiation by carotenoids in transformants of the cyanobacterium Synechococcus PCC7942. Plant Physiol 120:599–604

Grant WD, Tindall BJ (1986) The alkaline saline environment. In: Herbert RA, Codd GA (eds) Microbes in extreme environments. Academic, London, pp 25–54

Hagemann M, Erdmann N (1997) Environmental stresses. In: Rai AK (ed) Cyanobacterial nitrogen metabolism and environmental biotechnology. Springer, Heidelberg, pp 156-221 (Narosa Publishing House, New Delhi)

Hagemann M, Effmert U, Kerstan T, Schoor A, Erdmann N (2001) Biochemical characterization of glucosylglycerol-phosphate synthase of *Synechocystis* sp. strain PCC 6803: comparison of crude, purified, and recombinant enzymes. Curr Microbiol 43:278–283

Heising S, Schink B (1998) Phototrophic oxidation of ferrous iron by a *Rhodomicrobium vannieli* strain. Microbiology 144:2263–2269

Helm RF, Hung Z, Edward D, Leeson H, Peery W, Potts M (2000) Structural characterization of the released polysaccharide of desiccation tolerant *Nostoc commune* DRH- 1. J Bacteriol 184:974–982

Hill DR, Hladun SL, Scherer S, Potts M (1994) Water stress proteins of *Nostoc commune* (cyanobacteria) are secreted with UV-A/B absorbing pigments and associate with 1,4-b-D-xylanxylanohydrolase activity. J Biol Chem 269:7726–7734

Huber C, Wächtershäuser G (1997) Activated acetic acid by carbon fixation on (Fe, Ni)S under primordial conditions. Science 276:245–247

Javor BJ (1989) Hypersaline environments. Microbiology and biogeochemistry. Brock/springer series in contemporary bioscience. Springer, Berlin

Jiang H, Qiu B (2005) Photosynthetic adaptation of a bloom-forming cyanobacterium *Microcystis aeruginosa* (cyanophyceae) to prolonged UV-B exposure. J Phycol 41:983–992

Jones AC, Gu LC, Sorrels CM, Sherman DH, Gerwick WH (2009) New tricks from ancient algae: natural products biosynthesis in marine cyanobacteria. Curr Opin Chem Biol 13:216–223

Jungblut AD, Hawes L, Mountfory D, Hltzfeld B, Dietrich DR, Burns BP Neilan BA (2005) Diversity within cyanobacterial mat communities in variable salinity meltwater ponds of McMurdo Ice Shelf, Antarctica. Env Microbiol 7:519–529

Karsten U, Garcia- Pichel F (1996) Carotenoids and mycosporine- like amino acid compounds in members of the genus *Microcoleus* (Cyanobacteria): a chemosystematic study. Syst Appl Microbiol 19:285–294

Kehr JC, Gatte Picchi D, Dittmann E (2011) Natural product biosyntheses in cyanobacteria: a treasure trove of unique enzymes. Beilstein J Org Chem 7:1622–1635

Kovacik L (2000) Cyanobacteria and algae as agents of biodeterioration of stone substrata of historical buildings and other cultural monuments. In: Choi S, Suh M (eds) Proceedings of the New Millennium International forum on conservation of cultural property. Kongju National University, Kongju, Korea, pp 44–56

Kratochvil D, Volesky B (1998) Advances in the biosorption of heavy metals. Trends Biotechnol 16:291–300

Lemmon RM (1970) Chemical evolution. Chem Rev 70:95–109

Lesser MP, Stochaj WR (1990) Photoadaptation and protection against active forms of oxygen in the symbiotic prokaryote *prochloron*-sp and its ascidian host. Appl Environ Microbiol 56:1530–1535

Li Y, Lin Y, Loughlin PC, Chen M (2014). Optimization and effects of different culture conditions on growth of *Halomicronema hongdechloris*-a filamentous cyanobacterium containing chlorophyll f. Front plant sci 5:67

Mandal S, Rath J, Adhikary SP (2011) Adaptation strategies of the sheathed cyanobacterium *Lyngbya majuscula* to ultraviolet-B. Photochem Photobiol B Biol 102:115–122

Mansour HA, Shoman SA, Kdodier MH (2011) Antiviral effect of edaphic cyanophytes on rabies and herpes-1 viruses. Acta Biol Hung 62:194–203

Medina RA, Goeger DE, Hills P, Mooberry SL, Huang N, Romero LI Coibamide A et al (2008) A potent antiproliferative cyclic depsipeptide from the panamanian marine cyanobacterium *Leptolyngbya* sp. J Am Chem Soc 130:6324

Miller SR, Wingard CE, Castenholz RW (1998) Effects of visible light and ultraviolet radiation on photosynthesis in a population of the hot spring cyanobacterium, *Synechococcus* sp., subjected to high temperature stress. Appl Environ Microbiol 64:3893–3899

Miyake C, Michihata F, Asada K (1991) Scavenging of hydrogen peroxide in prokaryotic and eukaryotic algae: acquisition of ascorbate peroxidase during the evolution of cyanobacteria. Plant Cell Physiol 32:33–43

Miyashita H, Ikemoto H, Kurano N, Adachi K, Chihara M, Miyachi S (1996) Chlorophyll d as a major pigment. Nature 383:402–402

Murata N, Nishida I (1987) Lipids of blue-green algae (cyanobacteria). In: Stumpf PK (ed) The biochemistry of plants. Academic, San Diego, pp 315–347

Nägeli C (1849) Gattungen einzelliger Algen, physiologisch und systematisch bearbeitet Neue Denkschrift. AllgSchweiz Nat Ges 10:1–138

Neidhart FC, VanBogelen RA, Vaughn V (1984) The genetics and regulation of heat- shock proteins. Annu Rev Genet 18:295–329

Nishiyama Y, Yamamoto H, Allakhverdiev SI, Inaba M, Yokota A, Murata N (2001) Oxidative stress inhibits the repair of photodamage to the photosynthetic machinery. EMBO J 20:5587–5594

Nübel U, Garcia- Pichel F, Clavero E, Muyzer G (2000) Matching molecular diversity and ecophysiology of benthic cyanobacteria and diatoms in communities along a salinity gradient. Environ Microbiol 2:217–226

Nunnery JK, Mevers E, Gerwick WH (2010) Biologically active secondary metabolites from marine cyanobacteria. Curr Opin Biotechnol 21:787–793

Olson JM (2001) Evolution of photosynthesis (1970) rexamined thirty years later. Photosynth Res 68:95–112

Oren A (2000) Salt and brines. In: Whitton BA, Potts M (eds) The ecology of cyanobacteria: their diversity in time and space. 306. Kluwer Academic, Dordrecht, pp 281

Pang Q, Hays JB (1991) UV-B- inducible and temperature- sensitive photoreactivation of cyclobutane pyrimidine dimers in *Arabidopsis thaliana*. Plant Physiol 95:536–554

Pattanaik B, Adhikary SP (2002) Blue- green algal flora at some archaeological sites and monuments of India. Feddes Repert 113:289–300

Pattanaik B, Schumann R, Karsten U (2007) Effects of ultraviolet radiation on cyanobacteria and their protective mechanisms. In Algae and cyanobacteria in extreme environments. Springer Netherlands pp 29-45. (Excerpts used with permission)

Pereira AR, Cao ZY, Engene N, Soria-Mercado IE, Murray TF, Gerwick WH, Palmyrolide A (2010) An unusually stabilized neuroactive macrolide from palmyra atoll cyanobacteria. Org Lett 12:4490–4493

Pierson BK, Olson JM (1989) Evolution of photosynthesis in anoxygenic photosynthetic procaryotes. In: Cohen Y, Rosenberg E (eds) Microbial mats, physiological ecology of benthic communities. American Society of Microbiology, Washington, DC, pp 402–427

Post FJ (1977) The microbial ecology of the Great Salt Lake. Microb Ecol 3:143–165

Potts M, Friedmann EI (1981) Effects of water stress on cryptoendolithic cyanobacteria from hot dessert rocks. Arch Microbiol 130:267–271

Proteau PJ, Gerwick WH, Garcia- Pichel F, Castenholz R (1993) The structure of scytonemin, an ultraviolet sunscreen pigment from the sheaths of cyanobacteria. Experientia 49:825–829

Quesada A, Vincent WF, Lean DRS (1999) Community and pigment structure of Arctic cyanobacterial assemblages: the occurrence and distribution of UV- absorbing compounds. Fems Microbiol Ecol 28:315–323

Ramsing N, Prufert-Bebout L (1994) Motility of *Microcoleus chthonoplastes* subjected to different light intensities quantified by digital image analysis. Nato Asi Ser G Ecol Sci 35:183–183

Rath J, Adhikary SP (2007) Response of the estuarine cyanobacterium *Lyngbya aestuarii* to UV-B radiation. J Appl Phycol 19:529–536

Reed RH, Borowitzka LJ, Mackey MA, Chudak JA, Foster R, Warr SRC, Moore DJ, Stewart WDP (1986) Organic solute accumulation in osmotically stressed cyanobacteria. FEMS Microbiol Lett 39:51–56

Reynolds CS, Oliver RL, Walsby AE (1987) Cyanobacterial dominance: the role of buoyancy regulation in dynamic lake environments. NZJ Mar Freshwater Res 21:379–390

Richter FM, McKenzie DP (1981) On some consequences and possible causes of layered mantle convection. J Geophys Res 86:6133–6142

Ritter D, Yopp JH (1993) Plasma membrane lipid composition of the halophilic cyanobacterium *Aphanothece halophytica*. Arch Microbiol 159:435–439

Rudolph BR, Chandresekhar I, Gaber BP, Nagumo M (1990) Molecular modelling of saccharide-lipid interactions. Chem Phys Lipids 53:243–261

Scherer S, Potts M (1989) Novel water stress protein from a desiccation tolerant cyanobacterium. J Biol Chem 264:12546–12553

Scherer S, Ernst A, Chen TW, Böger P (1984) Rewetting of drought resistant blue- green algae: time course of water uptake and reappearance of respiration, photosynthesis and nitrogen fixation. Oecologia 62:418–423

Schiewer S, Volesky B (1996) Modeling of multimetal ion exchange in biosorption. Environ Sci Technol 30:2921–2927

Schilling JG (1973) Iceland mantle plume: geochemical study of Reykjanes ridge. Nature 242:565–571

Schwarzer D, Finking R, Marahiel MA (2003) Nonribosomal peptides: from genes to products. Nat Prod Rep 20:275–287

Shibata H, Baba K, Ochiai H (1991) Near- UV irradiation induces shock proteins in Anacystis nidulans R- 2; possible role of active oxygen. Plant Cell Physiol 32:77–776

Simmons TL, Nogle LM, Media J, Valeriote FA, Mooberry SL, Gerwick WH (2009) Desmethoxymajusculamide C, a cyanobacterial depsipeptide with potent cytotoxicity in both cyclic and ring-opened forms. J Nat Prod 72:1011–1016

Singh S, Kate BN, Banerjee UC (2005) Bioactive compounds from cyanobacteria and microalgae: an overview. Crit Rev Biotechnol 25:73–95

Sinha RP, Häder DP (1996) Response of a rice field cyanobacterium *Anabaena* sp. to physiological stressors. Env Exp Bot 36:147–155

Sinha RP, Klisch M, Häder DP (1999) Induction of a mycosporine-like amino acid (MAA) in the rice-field cyanobacterium *Anabaena* sp. by UV irradiation. J Photochem Photobiol B Biol 52:59–64

Sinha RP, Klisch M, Helbling EW, Häder D-P (2001) Induction of mycosporine-like amino acids (MAAs) in cyanobacteria by solar ultraviolet-B radiation. J Photochem Photobiol B Biol 60:129–135

Sompong U, Hawkins PR, Besley C, Peerapornpisal Y (2005) The distribution of cyanobacteria across physical and chemical gradients in hot spring in northern Thailand. FEMS Microbiol Ecol 52:365–376

Sørensen KB, Canfield DE, Oren A (2004) Salt responses of benthic microbial communities in a solar saltern (Eilat, Israel). Appl Environ Microbiol 70:1608–1616

Steinberg CEW, Schäfer H, Beisker W (1998) Do acid- tolerant cyanobacteria exist? Acta Hydrochim Hydrobiol 26:13–19

Tarko T, DuDa-ChoDak A, Kobus M (2012) Influence of growth medium composition on synthesis of bioactive compounds and antioxidant properties of selected strains of *Arthrospira* cyanobacteria. Czech J Food Sci 30:258–267

Teruya T, Sasaki H, Fukazawa H, Suenaga K (2009) Bisebromoamide, a Potent Cytotoxic Peptide from the Marine Cyanobacterium *Lyngbya* sp.: isolation, stereostructure, and biological activity. Org Lett 11:5062–5065

Tomo T, Okubo T, Akimoto S, Yokono M, Miyashita H, Tsuchiya T, Noguchi T, Mimuro M (2007) Identification of the special pair of photosystem II in a chlorophyll d dominated cyanobacterium. Proc Natl Acad Sci U S A 104:7283–7288

Tomo T, Allakhverdiev SI, Mimuro M (2011) Constitution and energetics of photosystem I and photosystem II in the chlorophyll d-dominated cyanobacterium *Acaryochloris marina*. J Photochem Photobiol B 104:333–340

Tripathi SN, Srivastava P (2001) Presence of stable active oxygen scavenging enzymes superoxide dismutase, ascorbate peroxidase and catalase in a desiccation-tolerant cyanobacterium *Lyngbya arboricola* under dry state. Current Sci 81:197–200

Tripathi A, Puddick J, Prinsep MR, Lee PPF, Tan LT (2010) Hantupeptins B and C, cytotoxic cyclodepsipeptides from the marine cyanobacterium *Lyngbya majuscula*. Phytochemistry 71:307–311

Tyystjärvi T, Herranen M, Aro E-M (2001) Regulation of translation elongation in cyanobacteria: membrane targeting of the ribosome nascent-chain complexes controls the synthesis of D1 protein. Mol Microbiol 40:476–468

Urrutia MM (1997) General bacteria sorption processes. In: Wase J, Forster C (eds) Biosorbents for metal ions. Taylor & Francis, London, pp 39–66

Vincent WF (2000) Cyanobacterial dominance in the polar regions. In: Whitton BA, Potts M (eds) The ecology of cyanobacteria. Kluwer Academic, Dordrecht, pp 321–340

Vincent WF, Quesada A (1994) Ultraviolet radiation effects on cyanobacteria: implications for Antarctic microbial ecosystems. In: Weiler CS, Penhale PA (eds) Ultraviolet radiation in Antarctica: measurements and biological effects Antarctic research series vol 62. American Geophysical Union, Washington, DC, 111–124

Volkmann M, Gorbushina AA, Kedar L, Aharon O (2006) Structure of euhalothece- 362, a novel red shifted mycosporine- like amino acid, from a halophilic cyanobacterium (*Euhalothece* sp.). FEMS Microbiol Lett 258:50–54

Wada H, Murata N (1990) Temperature-induced changes in the fatty acids composition of the cyanobacterium, Synechocystis PCC 6803. Plant Physiol 92:1062–1069

Walker JCG (1983) Possible limits on the composition of the Archaean ocean. Nature 302:518–520

Ward DM, Castenholz RW (2000) Cyanobacteria in geothermal habitats. In: Whitton BA, Potts M (eds) The ecology of cyanobacteria: their diversity in time and space. Kluwer Academic, Netherlands

Weckesser J, Broll C, Adhikary SP, Jorgensu J (1987) 2-O-Methyl-~-xylosec ontaining sheath in the cyanobacterium *Gloeothece* sp. PCC 6501. Arch Microbiol 147:300303

Welker M, Dohren HV (2006) Cyanobacterial peptides-nature's own combinatorial biosynthesis. FEMS Microbiol Rev 30:530–563

White RS, Brown JW, Smallwood JR (1995) The temperature of the Iceland plume and the origin of outward-propagating V-shaped ridges. J Geol Soc Lond 152:1039–1045

Wolfe- Simon F, Grzebyk D, Schofield O, Falkowski PG (2005) The role and evolution of superoxide dismutases in algae. J Phycol 41:453–465

Wu H, Gao K, Villafan~e V, Watanabe T, Helbling EW (2005) Effects of solar UV radiation on morphology and photosynthesis of the filamentous cyanobacterium *Arthrospira platensis*. Appl Environ Microb 71:5004–5013

Xiong S, Fan J, Kitazato K (2010) The antiviral protein cyanovirin-N: the current state of its production and applications. Appl Microbiol Biotechnol 86:805–812

Yakoot M, Salem A (2012) Spirulina platensis versus silymarin in the treatment of chronic hepatitis C virus infection. A pilot randomized, comparative clinical trial. BMC Gastroenterol 12:32

Yang H, Lee E, Kim H (1997) Spirulina platensis inhibits anaphylactic reaction. Life Sci 61:1237–1244

Ma Z, Gao K (2009) Photoregulation of morphological structure and its physiological relevance in the cyanobacterium *Arthrospira* (*Spirulina*) *platensis*. Planta 230:329–337

Zheng W, Chen C, Cheng Q, Wang Y, Chu C (2006) Oral administration of exopolysaccharide from *Aphanothece halophytica* (chroococcales) significantly inhibits influ enza virus (H1N1)-induced pneumonia in mice. Int Im munopharmacol 6:1093–1099

Chapter 2
Secondary Metabolites of Cyanobacteria and Drug Development

2.1 Secondary Metabolic Pathways Found in Cyanobacteria

Natural products, often called secondary metabolites, are low molecular weight organic molecules that have diverse and often very potent biological activities. Secondary metabolites are not essential for normal growth, development, or reproduction of an organism. They empower the producing organism to survive interspecies competition, provide defensive mechanisms against stress, and facilitate reproductive processes. Many secondary metabolites have proved invaluable as antibacterial or antifungal agents, anticancer drugs, cholesterol-lowering agents, immunosuppressants, antiparasitic agents, herbicides, diagnostics, and tools for research. Some of these have found to play a pivotal role in the treatment or prevention of a multitude of biological disorders. Many of the deadly diseases did not have any cure until these products were discovered. Secondary metabolites are commonly divided into structural classes related to their biosynthesis. This classification has its limitations because some compounds have building blocks from more than one biosynthetic pathway and some compounds that appear closely related can have completely different biosynthetic origins. The important classes of secondary metabolites are the polyketides and nonribosomal peptides, and other structural classes are alkaloids, terpenoids, shikimate-derived molecules, and amino glycosides. Recently, a new term "Parvome" has been proposed for these small molecules of great structural diversity (Davis and Ryan 2012).

Since the onset of the post-genomic era genomes of numerous microorganisms are constantly sequenced, and bioinformatic analysis continuously reveals a high number of biosynthetic pathways for the production of secondary metabolites, whereas only a few natural products can currently be correlated to the genome-sequenced strains. This discrepancy between the genetic capacity for secondary metabolite biosynthesis and low numbers of known compounds has fuelled the development of strategies aimed at the assignment of new secondary metabolites to the predicted genome-encoded pathways. Genome-mining needs, metabolome

© Springer International Publishing Switzerland 2015
S. Mandal, J. Rath, *Extremophilic Cyanobacteria For Novel Drug Development*,
SpringerBriefs in Pharmaceutical Science & Drug Development,
DOI 10.1007/978-3-319-12009-6_2

mining, and thus secondary metabolomics—inspired methods are of utmost importance for the success of genomics-based discovery of novel secondary metabolites.
Cyanobacteria are a rich source of natural products comprising of primary and secondary metabolites including nonribosomal proteins, polyketides and terpenes and
alkaloids, and several of these are known to have anticancer, antiviral, UV protective activities as well as hepatotoxicity and neurotoxicity (Herrero and Flores 2008).
Many of the cyanobacterial secondary metabolites have been characterized and are
known to be produced via the nonribosomal peptide synthetases (NRPS)/polyketide
synthases (PKS) route (Hoffmann et al. 2003). However, several other non-NRPS/
PKS synthetic pathways leading to the production of secondary metabolites with
the active participation of cytochrome P450 (CYP) are just beginning to be appreci-
ated, with the discovery of many CYP coding genes in their genome.

Many bioactive metabolites produced by cyanobacteria are either a peptide or
a macrolide structure, or a combination of both types (Welker and Von Döhren
2006). Other metabolites belong to the alkaloid class of compounds. Two types of
biosynthetic pathways produce the peptide class: by giant multidomain enzymes,
the NRPS or by ribosomal synthesis and subsequent post-translational modification and processing. NRPS consist of modules, each being responsible for the incorporation of a single amino acid. The order of these modules typically follows a
colinearity rule, i.e., the succession of modules corresponds to the order of amino
acids in the final product. A minimal module is composed of an amino acid-acti-
vating adenylation (A) domain, a peptidyl carrier proteins (PCP) domain carrying
the phosphopantetheine cofactor, and a condensation (C) domain. NRPS can accept
about 300 proteinogenic and nonproteinogenic substrates and may contain further
domains introducing tailoring modifications or epimerizing the amino acid substrates (Grünewald and Marahiel 2006). In contrast, ribosomal biosynthesis of peptides is limited to 20 proteinogenic amino acids. This group of peptides nevertheless
displays a high diversity and a considerable biosynthetic and bioactive potential.
The ribosomal pre-peptides are typically composed of a leader peptide and core
peptide. Associated post-translational modification enzymes (PTMs) catalyze different types of macrocyclizations of the core peptide and side chain modifications
of amino acids. Peptide maturation further requires cleavage of the leader peptide
by processing proteases (PP) frequently combined with transport across the plasma
membrane (Oman and Donk 2010). Macrolides in cyanobacteria are produced by
modular type PKS resembling NRPS with respect to their modular nature. In contrast to the peptide-synthesizing enzymes, in PKSs different types of carboxylic
acids are activated, assembled, and optionally modified. The maximal set of domains of an individual PKS module is identical to animal fatty acid synthase (FAS)
(Jenke-Kodama et al. 2005) and consists of ketosynthase (KS), acyltransferase
(AT), ketoreductase (KR), dehydratase (DH), enoyl reductase (ER) and acyl carrier
protein (ACP) domains (Staunton and Weissman 2001). Parts of the domains (KR,
DH, ER) are optionally used leading to a different reduction state of the keto groups
of polyketides. There are also alternative PKS assembly lines cooperating with AT
domains encoded in trans of the multi enzymes (Piel 2010), or PKS types comprising single modules that work iteratively (Campbell and Vederas 2010).

The major trait of cyanobacterial pathways is their hybrid character, i.e., the frequent mixture of NRPS and PKS modules. The first biosynthetic pathway identified and partially characterized for cyanobacteria was the mixed NRPS/PKS pathway catalyzing the formation of the hepatotoxin microcystin in the cyanobacterium *Microcystis aeruginosa* (Tillett et al. 2000). The pentapeptide nodularin is structurally closely related and shares a highly similar biosynthetic pathway (Moffitt and Neilan 2004). The anabaenopeptilide pathway in the strain *Anabaena* 90 was described soon after first reports about microcystin biosynthesis. Anabaenopeptilides belong to the cyanopeptolin family of depsipeptides that were shown to inhibit different types of serine proteases (Welker and von Döhren 2006). The signature of this group is the unusual 3-amino-6-hydroxy-2-piperidone moiety (Ahp). The corresponding NRPS assembly line consists of seven modules (Rouhiainen et al. 2000). Unique features of anabaenopeptilides include an integrated formyl transferase domain in the initiation module and nicotinamide adenine dinucleotide dependent (NAD-dependent) halogenase. Aeruginosins are specific inhibitors of serine type proteases and produced by different genera of freshwater cyanobacteria. *Planktothrix agardhii* produces glycosylated variants of the peptides, aeruginosides, via a mixed NRPS/PKS pathway. The signature of this group is the 2-carboxy-6-hydroxyoctahy-droindole (Choi) moiety. The loading module was predicted to activate phenyl-pyruvate, which is reduced by an integrated KR domain to phenyllactate. Several NRPS/PKS assembly lines were identified and partially characterized for the marine cyanobacterium *Lyngbya majuscula* (Chang et al. 2002). The first pathway described was the biosynthesis of barbamide, a chlorinated lipopeptide with potent molluscidal activity. The lipopeptide contains a unique trichloroleucyl starter unit that is halogenated by unique biochemical mechanisms through the two nonheme iron (II)-dependent halogenases BarB1 and BarB2. Further extraordinary features of the pathway include one-carbon truncation during chain elongation, *E*-double bond formation, and thiazole ring formation. The other important secondary metabolites produced by NRPS/PKS pathways are aeruginosin, cylindrospermopsin, anatoxin, jamaicamide, curacin A, hectochlorin, lyngbyatoxin, apratoxin, nostopeptolide, nostocyclopeptide, cryptophycin.

The majority of cyanobacterial peptides are produced nonribosomally, specifically two peptide families, namely patellamides and microviridins, for which no NRPS pathway could be assigned. Genome-scale analyses have unrivaled further peptide families. Cyanobacteria can now be considered as one of the most prolific sources of ribosomal-produced natural products. The first ribosomal pathway discovered was the biosynthesis of patellamides in the symbiotic cyanobacterium *Prochloron*. The cyclic octapeptides are pseudosymmetric and contain thiazole and oxazolin rings. Patellamides are typically moderately cytotoxic, and some variants were further reported to reverse multidrug resistance (Schmidt et al. 2005). Microviridins are a group of tricyclic depsipeptides predominantly detected in bloom-forming freshwater cyanobacteria. Several members of the family potently inhibit various serine-type proteases. The biosynthetic pathway of microviridins was described for the genera *Microcystis* and *Planktothrix* (Ziemert et al. 2008). Posttranslational modification of microviridins is achieved by the activity of two closely related adenosine triphosphate (ATP) grasp ligases, MdnB and MdnC. The enzymes introduce two ω-ester linkages between threonine and aspartate and serine and glutamate (MdnC/MvdD) and one ω-amide linkage between lysine and aspartate (MdnB/MvdC). Cyclizations occur in a strictly defined order. Ring size and composition of the microviridin core peptide is invariant (Philmus et al. 2009), whereas N-terminal and C-terminal amino acids are highly variant (Ziemert et al. 2010). The enzyme system further contains a GNAT-type N-acetyltransferase and an ABC transporter.

Cyanobacteria produces two types of sunscreen compounds induced under UV irradiation: Scytonemin and mycosporine-like amino acids. Biosynthesis of the two groups of compounds has recently been elucidated, providing further examples for the fascinating natural product biochemistry of cyanobacteria. A gene cluster responsible for scytonemin biosynthesis was initially discovered by random mutagenesis in the terrestrial symbiotic cyanobacterium *Nostoc punctiforme* (Soule et al. 2007). The gene cluster contains a number of genes related to aromatic amino acid biosynthesis (Soule et al. 2007). The biosynthetic route was proposed to start with tryptophan and tyrosine. Two of the initial steps of the sunscreen synthesis were reproduced *in vitro* (Balskus and Walsh 2008). The open reading frames (ORF) NpR1275 was confirmed to act as a tryptophan dehydrogenase, whereas p-hydroxyphenylpyruvic acid was proposed to be generated by the putative prephenate dehydrogenase NpR1269. Both substrates are then further transformed by the thiamin diphosphate (ThDP)-dependent enzyme NpR1276 to isomeric acyloins representing one-half of the carbon frameworks of scytonemin (Balskus and Walsh 2008). The enzyme showed a remarkable selectivity for the specific C–C bond reaction that is unprecedented in natural systems. The other sunscreen compound microsporines consist of a single amino acid linked to cyclohexenone. Cyanobacteria and other algae produce mycosporine-like amino acids, which contain two substituents linked to the central ring by imine linkages. Four enzymes are involved in the synthesis of the specific MAA (mycosporine-like amino acid) shinorine in *Anabaena vari-abilis* ATCC 29413: a dehydroquinase synthase homologue (DHQS), an O-methyl-transferase (O-MT), an ATP grasp ligase, and an NRPS-like enzyme.

There is considerable evidence that some stress responses can be used to trigger the expression of secondary metabolic genes. Altering a single parameter in the growth conditions and eliciting a stress response has been previously applied through the one strain many compounds (OSMAC) approach to explore the secondary metabolic potential of different strains of cyanobacteria (Edwards and Ericsson 1999). Cytochrome P450s CYPs are a group of ubiquitous hemoprotein oxygenases that are found in all domains of life (Nebert et al. 1989). The great diversity, occurrence, and distribution of CYP suggest that they could be involved in essential or crucial metabolism, such as defense against environmental pollutants, drug detoxification, synthesis of important molecules, and defense against extreme environmental (Bernhardt 2006). Several CYP and CYP-like genes have been identified in cyanobacterial genomes, however, little attention has been given to their functional characterization (Ke et al. 2005; Kühnel et al. 2008; Alder et al. 2009). Elucidation of the crystal structure of CYP120A1 has shown that it participates in retinoid metabolism (Ke et al. 2005). A BLASTp search in the National Center for Biotechnology Information (NCBI) cyanobacterial genome database using the characterized protein NP_488726 from *Nostoc* sp. strain PCC 7120 (Agger et al. 2008) or CYP120A1 from *Synechocystis* sp. PCC 6803 (Kühnel et al. 2008) yields 100 putative CYP sequences and it distributed in most of the known cyanobacterial species. CYP110 from *Nostoc* sp. PCC 7120 is not a cytosolic enzyme, but membrane bound, like eukaryotic CYPs. It is not induced by alkanes, and does not participate in alkane biodegradation, but it is involved in v-hydroxylation of long chain fatty acids and plays a role in nitrogen fixation (Torres et al. 2005). Isoprenoids are one of the major structural classes of natural products. Recent efforts have shown that cyanobacterial strains are capable of producing isoprenoids (Agger et al. 2008). In addition, marine cyanobacteria from *Microcoleus* and *Phormidium* genera have been known to contribute to bioremediation of oil spills using CYP as a catalyst for the alkane breakdown (Hasan et al. 1994). Though cyanobacteria are important as a source of natural product the CYP related pathways are only recently becoming apparent. The increasing number of functionally characterized CYPs from cyanobacteria, as well as evidence of terpene synthase genes, is opening new vistas of natural product formation (Agger et al. 2008).

While some studies attempted the activation of biosynthetic pathways which were assumed as being "silent" under standard laboratory conditions, through the variation of cultivation methods, by introducing environmental challenges or by modifying regulatory mechanisms (Hertweck 2009) others aimed to provide evidence at the transcriptomic and proteomic level to help the identification of active biosynthetic pathways and subsequently uncover the corresponding metabolites (Schley et al. 2006). As a straightforward method to directly assign novel metabolites to predicted biosynthetic pathways, the construction of knockout mutants by targeted gene inactivation coupled to comparative metabolite profiling has been used successfully to disclose previously undiscovered secondary metabolites (Cortina et al. 2012). Furthermore, nowadays a variety of methods have been developed that follow a holistic "metabolites first" principle, i.e., all metabolites in a given analytical method are able to detect are first recorded, and using of hyphenated high-performance liquid chromatography in combination with high-resolution tandem mass spectrometry (LC-HRMS), and matrix assisted laser desorption/ionization imaging mass spectrometry (MALDI-IMS) were analyzed in a high throughput scheme.

2.2   Recent Advances in Novel Secondary Metabolites Discovery from Cyanobacteria

With advances in analytical techniques, there is a tremendous increase in discoveries of secondary metabolites. Low cost, chemical free green extraction (GE) methods such as supercritical fluid extraction (SFE), pressurized liquid extraction (PLE) are now gaining popularity for extraction of secondary metabolites. Analytical techniques like advanced ultra-performance liquid chromatography (UPLC), which can be a better option than high-performance liquid chromatography (HPLC). Many cyanobacteria can be cultured as single-species communities. The microbial communities, or colonies, curate their environment via metabolic exchange factors such as released natural products. To date, there are very few tools available that can monitor, in a systematic and informative fashion, the metabolic release patterns by microbes grown in a pure or mixed culture. There are significant challenges in the ability to monitor the metabolic secretome from growing microbial colonies. For example, the interactions of such molecules can be extremely diverse, ranging from polyketides, nonribosomal peptides, isoprenoids, fatty acids, and microcins to peptides, poly-nucleotides, and proteins (Cane 2010). Because of this chemical diversity, most of these molecules are extracted prior to analysis and studied one at a time and apart from the native spatial context of a microbial colony. Thus, limited information is obtained about the metabolic output of colonies in a synergistic or multiplexed fashion. Genomics is the most prominent tool to define the makeup and species dynamics of the microbiome. Once the complete genome of an organism is sequenced, genome-mining approaches can be used to predict natural products and to discover novel adaptive metabolites. Generation of annotated genomes helps to provide a basis for predictive algorithms capable of mining unannotated genomes for new secondary metabolites, which can drastically improve the speed at which new molecules can be targeted and characterized. We discuss here recent advances in the extraction of novel secondary metabolites from cyanobacteria.

2.2.1 Epigenetic-Related Approaches

Secondary metabolites in cyanobacteria confer an evolutionary benefit to the producing organism. In the simplified environment of the laboratory, cyanobacteria often do not depend on the entire capabilities of their secondary metabolome and thus the products of most of the biosynthetic gene clusters are not observed. Improvements in *de novo* genome sequence technologies have resulted in a dramatic increase in the number of complete genomes available for well-known producers of natural products. These data have revealed that many members of these groups produce only a small fraction of the natural products encoded in their genomes under standard laboratory conditions. The natural product biosynthetic pathways that are not expressed, often referred to as the "silent metabolome," therefore, potentially represent a vast reservoir of undiscovered small molecules. Epigenetic enzymes like histone deacetylases (HDACs) and DNA methyltransferases (DNMTs) play a crucial role in gene regulation of biosynthesis clusters (Schmitt et al. 2011). Recently, it was discovered that interference with these systems can result in upregulation of biosynthetic clusters (Cichewicz 2010). In chemical epigenetics, HDAC and DNMT inhibitors are used to manipulate the epigenome. Similar effects can be obtained through genetic manipulation of genes encoding global transcriptional regulators like LaeA (Bok et al. 2006), histone deacetylases (Lee et al. 2009), or methylating enzymes (Bok et al. 2009), as well as the small ubiquitin-like modifier (SUMO) protein (Szewczyk et al. 2008). The complexity of the affected systems is demonstrated by the fact that, in all cases, gene deletion and over expression increased the expression of some biosynthetic clusters and decreased the expression of others. In this context, epigenetic approach is a novel tool to get valuable metabolites from cyanobacteria.

2.2.2   Genome-Driven Approaches

Genome mining as an approach to natural product discovery has recently been studied (Gross 2009; Velasquez andDonk 2011). While it is generally possible to identify the biosynthetic gene cluster for a known compound produced by a microorganism from genome sequence data, the converse approach of predicting the exact structure of a natural product from sequence data is often not possible. Several factors contribute to this problem, including difficult to predict post-assembly, modification, biosynthetic domain skipping, ambiguous cyclization patterns, and noncolinearity of some biosynthetic enzymes. Although bioinformatics tools exist to analyze genome data, to identify natural product biosynthetic clusters with a low level of accuracy, to predict the structure of the encoded compound, there is room for significant advancement in this field. There are possibility to identify silent gene clusters in natural product producing microorganisms by subtractive analysis, comparing the observed compounds to biosynthetic pathways predicted using existing bioinformatics tools (Bachmann and Ravel 2009; Schmitt et al. 2011). Annotated genome sequence data can facilitate directed genetic strategies aimed at de-silencing individual gene clusters. Positive regulators are often colocalized with the biosynthetic pathways under their control. Integration of a functional promoter in front of these regulators has the potential to upregulate the entire biosynthetic pathway. This approach requires a method to introduce DNA into the genome of the microorganism via homologous recombination, as well as rudimentary genetic tools including a validated promoter and a suitable selectable marker. The same strategy can be used to integrate a functional promoter in front of a silent biosynthetic pathway; although this strategy may only be practical when a single transcript is predicted to contain all of the necessary genes for production of the compound. Expression of silent biosynthetic pathways in heterologous hosts offers another strategy for de-silencing biosynthetic pathways. The emergence of phage protein based recombining approaches has significantly aided in manipulating large DNA fragments and has made heterologous expression of natural product pathways a viable option (Thomason et al. 2007). Heterologous expression of pathways cloned directly from DNA isolated from environmental samples (a culture independent approach) for accessing natural product chemical diversity, is under explored (Feng et al. 2010). Currently, heterologous expression is a very labor-intensive strategy that limits its broad application in the drug discovery process; however, when a biosynthetic pathway of high interest has been identified this approach offers an additional tool to study the pathway in a genetically tractable host.

Genome mining has long been used to predict molecular structures (Liu et al. 2010) and is still used to successfully characterize novel NRPS, PKS, terpenoid, and other natural products. Genome-mining searches for genes or gene clusters that encode enzymes involved in the biosynthesis of natural products are based on se-quence alignment with other characterized enzymes involved in that specific natural product biosynthesis. Genome-screening programs such as ClustScan, NRPS-PKS, and "NP. Searcher" are used to predict locations of gene clusters and the structure of their putative products (Challis 2008). Success of genome mining depends on the availability of complete microbial genomes, thus will prove especially powerful when used in conjunction with metagenomics, as sequenced genomes and plasmids can be automatically fed into a pipeline to be mined for secondary metabolite pro-duction (Challis 2008). The tools currently in place provide a good starting point for data mining; however, improvements are still needed in predictive software for adaptive metabolites, including a ribosomal encoded peptide predictor, incorpo-ration of 6-frame translation into genome-mining searches, and consolidation of metabolite and small molecule databases.

2.2.3 Analytics and Preparative Strategies for Accessing New Natural Products

Though many tools contribute to "-omics" studies and natural product workflows, mass spectrometry (MS) and nuclear magnetic resonance (NMR) are the two most powerful tools. They are already used for individual "-omics" approaches. MS is ca-pable of high throughput identification of proteins, metabolites, and other types of molecules through the generation of important structural information with tandem MS. Like MS, NMR is often used for compound identification as well as to observe global metabolite changes. NMR is able to provide structural information, atomic connectivity, and stereochemistry that MS cannot. Ongoing improvements of these two technologies have allowed for optimal data generation from a small sample size or from crude samples. Nanomolar NMR elucidates structures from as little as one nano mole of material (Molinski 2010), although it is anticipated that this will be in the picomolar range in the near future, while some MS methods can evaluate sam-ples in the sub-picomolar range. NMR of crude mixtures has been previously used for a variety of samples and is advantageous since it eliminates the need to separate molecules prior to analysis (Exarchou et al. 2005; Schroder et al. 1998). Combin-ing liquid chromatography (LC) with MS or NMR allows for the separation of molecules preceding analysis, resulting in improved signal intensity. Additionally, spatial localization of molecules can be characterized using IMS (Imaging mass spectrometry) (Cornett et al. 2007), often giving important information about signal distribution in the cell (Seeley and Caprioli 2008) or in interactions grown on agar (Yang et al. 2009). Each tool has an invaluable role in "-omics" studies and has great potential in advancing microbiome-based workflows in studies that will both catalog known molecules and target uncharacterized molecules.

A breakthrough in natural product analytics has been the introduction of ultrahigh pressure liquid chromatography (UHPLC) (Wolfender et al. 2010). In combination with time-of-flight (TOF) MS detectors with very fast response times, UHPLC is a very efficient tool for de-replication of natural product extracts and for genome-driven identification of new natural products (Grata et al. 2008). The excellent sensitivity and resolution of contemporary chromatography systems enables the miniaturization of the entire process of broth screening making it much more powerful and less cumbersome than in the past. In particular, principal component analysis (PCA) has become increasingly popular. The main concept of PCA is to reduce a large dataset, obtained for example from different LC–MS experiments, in order to extract the most important variations between the samples without significant loss of information (Kuhnert et al. 2011). For structure elucidation of these compounds, LC–NMR spectroscopy can be used. As on-flow or stop-flow LC–NMR methods having inherently low sensitivity and are limited to the measurement of proton NMR spectra, SPE-NMR (trapping of the LC-flow on a SPE [Solid-phase extraction] column) or CapNMR (micro fractionation followed by measurement of a concentrated sample via micro-flow capillary LC–NMR probes) can overcome these limitations and can be used for structure elucidation through 1D and 2D measurements (Wolfender et al. 2010). For the purification of natural products, supercritical fluid chromatography (SFC) is becoming a powerful tool. While this technology was introduced nearly 50 years ago, only in recent years have many of the instrumental limitations been overcome. In combination with MS detection, new natural products can be isolated very efficiently using MS-guided fractionation (Guiochon and Tarafder 2011). All these technical innovations have reshaped natural product chemistry significantly over the past couple of years. The reisolation of known secondary metabolites could be avoided early in the process using high-resolution analytical tools. The bothersome purification of natural products can be performed semiautomatically and much more efficiently with the current instrumentation. As a consequence of these technical improvements the costs dropped to generate pure natural product libraries.

The relatively rapid execution and chemically specific information provided by MALDI IMS experiments offers an enticing tool to enhance the preclinical trials of pharmaceutical compounds (Gessel et al. 2014). For a targeted approach, using the known molecular mass of a compound offers a distinct advantage in defining its localization, allowing experimenters to easily distinguish the target from unknown species and/or electronic and chemical noise. Proteomic changes arising from treatment with a specific pharmaceutical compound can be tracked in conjunction with the distribution and metabolism of the compound. Imaging of pharmaceutical compounds and their metabolites is quite challenging because of the small size of the drug molecules, but with high resolution mass analyzers these species can be successfully isolated in the lower mass regions (100–500 Da) (Chughtai and Heeren 2010). In a recent study Shanta and colleagues addressed the issue of matrix interference in low mass regions and drug/drug metabolite detection via MALDI IMS

(Shanta et al. 2012). They detail the development of a binary matrix compound that has significantly reduced interference peaks from 0 to 500 Da. IMS is a powerful tool for simultaneously investigating the spatial distribution of multiple different biological molecules. The technique offers a molecular view of the peptides, proteins, polymers, and lipids produced by a microbial colony without the need for exogenous labels or radioactive trace material (Gonzalez et al. 2012). Target compounds can be measured and visualized simultaneously and in a high throughput manner within a single experiment. IMS extends beyond techniques such as MALDI profiling or MALDI intact cell analysis. Although invaluable, these techniques give a broad view of the metabolites produced in reference to a growing colony, where discretely secreted low global concentration but high local concentration metabolites could be missed. IMS entails examining the entire cyanobacterial colony, including the surrounding agar medium, by defining a raster composed of greater than 1000 laser spots (points of data collection), which increases the likelihood of detecting unique, discrete ion distribution patterns, and hidden molecular phenotypes that cannot be observed by the naked eye. Using MALDI-TOF, several new cyanopeptides were observed and characterized. Recently, MALDI-TOF was used to directly analyze cyanobacteria for the presence of cyanopeptides and toxins from 850 individual colonies (Welker et al. 2006). In this study, a small colony of the cyanobacterium was placed on a MALDI-TOF-plate and covered with a small amount of 2,5-dihydroxybenzoic acid matrix before they were analyzed by MALDI-TOF mass spectrometry. A total of 90 individual peptides was identified from these 850 individual cyanobacteria colonies, including 18 that appear to be unique from their masses. Erhard et al. (1997) used MALDI-TOF MS for identification of secondary metabolites with intact cyanobacterial cells. Resulting mass signals, which are further characterized by post source-decay fragmentation, and comparison of observed fragment spectra with theoretical ones or with those of pure reference compounds. Desorption electrospray ionization mass spectrometry (DESI-MS) is another applied analytical technique for chemical profiling, characterization and quantification of low molecular weight biomolecules (Esquenazi et al. 2008). Another technique is a direct analysis in real-time mass spectrometry (DART-MS), which is very much effective in chemical profiling and fingerprinting of bioactive molecules without prior sample preparation. Singh and Verma (2012) have identified the *Nostoc* sp. on the basis of characteristic chemical compounds (chemical finger printing) using DART-MS. Advances in all these analytical techniques have a tremendous impact on the identification and characterization of natural products and hope many new bioactive natural products will come up with advances in this area.

2.3   *In silico* Screening for Novel Secondary Metabolites of Cyanobacteria

Traditionally, drug-target discovery is fundamentally a wet lab experimental process comprising of identification of candidate lead compounds in chemical/activity-guided screening programs, where extracts and purified compounds are tested against specific targets or in whole cell assays. Introduction of new drugs and novel therapeutic solutions is a long and costly process (DiMasi et al. 2003). Pharmacologists strive to optimize and accelerate this process by developing new *in vivo* and *in vitro* investigation strategies. Genetic screening has complemented these approaches by applying DNA probes of conserved biosynthetic enzymes to identify the biosynthesis gene clusters in gene libraries of the producing organisms. While the underlying principles behind *in vitro* drug-target discovery have not changed much in the past, there are profound changes in the throughput or speed by which this process is done. In the past, small numbers of drugs and drug targets were slowly identified through manually intrusive and exceedingly tedious laboratory processes. Nowadays, large numbers of potential drugs and drug targets are being routinely identified through a variety of high speed, robotized technologies, including high throughput DNA sequencing (Kramer and Cohen 2004), high throughput microarray or two-dimensional (2D) gel electrophoresis experiments (Onyango 2004), rapid-throughput mass spectrometry assays, and high-speed robotized chemical library screens (Comess and Schurdak 2004). Both, chemical and genetic screening is very labor- and time intensive and therefore costly. With the progress in sequencing technologies, *in silico* screening was developed as a third method, which greatly speeds up the discovery of novel compounds. *In silico* approach uses genetic/genomic information to assess the genetic potential of microorganisms for the ability to produce novel compounds (Walsh and Fischbach 2010). The majority of the *in silico* methods are primarily used in parallel with the generation of *in vivo* and *in vitro* data for accurate modeling and validation of a wide range of applications from the ligand design and optimization to the characterization of fundamental pharmacological properties of molecules such as absorption, distribution, metabolism, excretion, and toxicity (Ekins et al. 2007). The diversity of the developed mathematical and biophysical models in this field resembles the manifolds of the pharmacological problems uniquely.

    In silico drug-target discovery is now possible primarily because of the Human Genome Project (Hopkins and Groom 2002) and related large scale sequencing efforts. Already more than 2000 viral genomes, 260 bacterial genomes, and more than two dozen eukaryotic genomes have been sequenced and deposited into public databases (http://www.ebi.ac.uk/genome). These data are allowing researchers to identify literally thousands of drugs for both endogenous diseases (Central nervous system disorders, diseases of aging, autoimmune diseases, and acquired or inborn metabolic disorders) and infectious diseases (bacterial, viral, and parasitic diseases). Many of these targets are being, or can be, rapidly identified *in silico*, using simple sequence comparison and sequence alignment software. Prior to these large-scale genomic sequencing efforts, the total number of endogenous (human) disease genes targeted by all existing drugs was estimated to be less than 400 (Hopkins and Groom 2002). Now, with all the sequence data in hand, it is estimated that the number of viable endogenous disease drug target could grow from ~300 to at least 3000 and the number of viable infectious disease drug target or drug-target classes could grow from ~20 to at least 300 (Hopkins and Groom 2002). Most of the drug targets can be either large molecules (protein, DNA, and RNA) or small molecules (metabolites). Ninety-six percent of approved drug target types are peptides or protein molecules with 93% of all nonredundant US FDA approved drugs being targeted to proteins. Even if a druggable protein or set of proteins is identified, there is an added challenge of finding a drug to make the drug target(s) functional—while not adversely affecting other normal functions of the body. This is a difficult task, especially for a small molecule.

The first step of the *in silico* approach is to identify putative biosynthetic gene clusters in the genome sequence. In a second step, the genetic information is then used to predict the biosynthetic products synthesized by the enzymes encoded in the gene clusters. To perform these tasks, specialized software tools are required. There are three commonly used tools SMURF, antiSMASH, and NP.searcher used for this screening (Weber 2014). The raw material for *in silico* drug target discovery is generally sequenced data, either a protein or nucleic acid sequence. There are more than 73 cyanobacterial genomes are either completely sequenced, or in draft or in near completion phase. The next step in this process is to identify genes that may be a part of a system leading to the discovery of potential therapeutically useful compounds. *In silico* analysis of cyanobacteria have attracted increasing attention only recently owing to their potential in the production of pharmaceuticals and other bioactive compounds. Shastri and Morgan (2005) used flux balance analysis for the investigation of *synechocystis* sp. PCC 6803 metabolism.

Identification of new biologically active compounds is required for development of new drugs. *In silico* approach have the potential not only speeding up the drug discovery process, thus reducing the costs, but also about changing the way drugs are designed. Rational Drug Design (RDD) helps to facilitate and speed up the drug designing process, which involves a variety of methods to identify novel compounds. One such method is the docking of the drug molecule with the receptor (target). The site of drug action, which is ultimately responsible for the pharmaceutical effect, is a receptor. Docking is the process by which two molecules fit together in 3D space. Nowadays, molecular docking approaches are routinely used in modern drug design to understand the drug-receptor interaction. Computer-aided drug design uses computational chemistry to discover, enhance, or study drugs and related biologically active molecules. The most fundamental goal is to predict whether a given molecule will bind to a target and if so how strongly. Molecular dynamics are most often used to predict the conformation of the small molecule and to model conformational changes in the biological target that may occur when the small molecule binds to it (Rajamani and Good 2007). In docking, a large number of potential ligand molecules are screened. This method is usually referred as ligand-based drug design (Kroemer 2007) and the nature of the complex between the drug and the receptor complex can be identified via docking and their relative stabilities for inhibition can be evaluated using molecular dynamics and their binding affinities using free energy simulations.

Metabolites of cyanobacteria are an excellent source of docking studies. More than 50% of the marine cyanobacteria are potentially exploitable for extracting bioactive substances, which are effective in either killing the cancer cells by inducing apoptotic death. In a recent study using molecular docking it has been found that the cyanobacterial drug Cryptophycin F isolated from *Nostoc* sp., is very potential against breast cancer causing receptor ERα in comparison to two commercial drugs Toremifene and Raloxifene (Sangeetha et al. 2014). Among the various members of marine cyanobacteria, *Calothrix, Lyngbya* sp., *Lyngbya confervoides, Lyngbya majuscule, Lyngbya sordida, Nostoc* sp., *Phormidium gracile, Symploca* sp., and *Symploca hydnoides* are highly potential organisms having anticancer drug molecules such as antillatoxin 2, apratoxin B1, arulide 3, baslynbiyaside, belamide A2, basibroamide 1, calothrixin B, caylobolide A2, cryptophycin 226, kemopeptinde B, hoamide C, homodolastin, isomalgamide B, lagunamide C, lynbyabelin C, lynbaysolide B1, lyngbastatin, majusculamide D, maleviamide D, malyngamide R, malyngolide, nostocylopeptide, obynanaide, pitipeptolide D, pitiprolamide, pompanopeptin, somocystinamide A, symplocamide A, symplostatin, tasipeptin B, tasiamide B, tiglicamide A, and veraguamide F. When these drug molecules were docked with the lung cancer causing receptor molecule EGFR (Epidermal growth factor receptor) kinase, Tiglicamide A showed a maximum Glide score indicating effective molecules against receptor tumor causing molecule (Vijayakumar and Menakha 2014). In another study, it is concluded that 31 bioactive compounds were isolated from 8 species of cyanobacteria as drugs for various types of cancers. Among the 31 bioactive compounds, tasiamide-B from *Symploca* sp was identified as the best drug for skin cancer compared with the commercially available drugs, like cabazitaxel and dyclonine. Through molecular docking, tasiamide-B was found to be more effective by interacting strongly with skin cancer causing target protein, HSP90 with the help of Glide module (Schrodinger suite; Vijayakumar and Menakha 2013). All these studies show with the rapid progress of *in silico* approaches, it could be expected that biomedical investigations into virtual reality ultimately lead to rigorous changes in the pharmaceutical research landscape by optimizing the drug development process, reducing the number of animal experiments and smoothing the path to personalized medicine. With the growing understanding of complex diseases, the focus of drug discovery has shifted away from the well-accepted "one target, one drug" model, to a new "multi-target, multi-drug" model, aimed at systemically modulating multiple targets. Identification of the interaction between drugs and target proteins plays an important role in genomic drug discovery, in order to discover new drugs or novel targets for existing drugs. Due to the laborious and costly experimental process of drug-target interaction prediction, *in silico* prediction could be an efficient way of providing useful information in supporting experimental interaction data. An important notion that has emerged in post-genomic drug discovery is that the large-scale integration of genomic, proteomic, signaling, and metabolomic data can allow the construction of complex networks of the cell that would provide with a new framework for understanding the molecular basis of physiological or pathophysiological states required for multi-target multi-drug discovery

2.4 Novel Biosynthetic Gene Cluster in Cyanobacteria

While identification of secondary metabolites from cyanobacteria has a long history, the identification of the biosynthetic genes responsible for the production of secondary metabolites in them is relatively recent. Among the 600 cyanobacterial metabolites isolated there are only 30 cases for which the biosynthetic genes have been identified. The main reason for this low number of gene clusters discovery is that the identification of these genes is highly dependent on genomic data. There are about 73 cyanobacterial genomes so far sequenced and the majority of them are from unicellular species that do not produce much secondary metabolites (Hess 2011). Further the genetic manipulations of cyanobacteria are difficult and probably hamper the discovery of biosynthetic genes implicated in secondary metabolism.

The majority of the metabolites of cyanobacteria are products of PKS, NRPS, or more frequently of PKS/NRPS hybrids, but several recent studies suggest they RPs (Ribosomal Peptide synthetase) are much more common in cyanobacteria (Xu et al. 2011). However, the PKSs and NRPSs are capable to produce very different molecules with diverse biological activities, even though their basic functions supported by the individual domains are conserved. PKSs and NRPSs are large, multifunctional protein complexes that catalyze the stepwise condensation of simple metabolic building blocks. Both PKSs and NRPSs have a modular organization, with each module carrying all essential information for the recognition, activation, and modification of one substrate (amino acids for NRPS or coenzyme A thioester derivatives of carboxylic acids for PKS) into the growing chain. Each module can be further divided into different domains, each responsible for a specific biochemical reaction. The number of modules and their domain organization within the enzymes control the structures of the final products (Schwarzer et al. 2003). The biosynthetic gene clusters in cyanobacteria are quite large, over 20 kb for PKs, NRPs and PK/NRP hybrid and about 10 kb in case of RPs since in this case the biosynthetic machinery is the ribosome. In case of scytonemin, alkanes, alkenes, and terpene the clusters are very small in size. The identification of biosynthetic gene clusters in cyanobacteria followed the same scenario in many cases. The structure of the metabolites was characterized followed by the feeding experiments using isotopically labeled precursors, giving some clues concerning the biosynthesis and a particular gene was then identified by using degenerative polymerase chain reaction (PCR) amplifications. The link between the identified gene and the biosynthesis was then inferred from two different types of data. Then the genetic inactivation of the identified gene abolished the production of the metabolites by unambiguous preferred experiments or the correlation between the presence of the gene (genotype) and the production of the metabolites (phenotype). Then, using either genomic libraries cloned in the cosmos or genomic sequence data, the clusters were identified and sequenced. In some cases, genome mining directly identified gene clusters. Several gene clusters responsible for the biosynthesis of the same metabolites have been sequenced in different strains or genus. For example, *mcy*, *stx*, *cyr*, *are*, *ana*, and *cyanobactin* gene clusters. The results so far obtained indicate that horizontal gene transfer is responsible for the presence of the same gene clusters in the genome of cyanobacteria of different genera. Another general trait is that usually one cluster is responsible for the biosynthesis of several variants of one metabolites. The variations observed are either a change in amino acid residue, a methylation, a hydroxylation, a sulfatation, a halogenation, or a change in configuration at some carbons. These variations are either catalyzed by tailoring enzymes or by some domains of the PKSs or the NRPSs that show a relaxed specificity.

Microcystins are nonribosomal cyclic heptapeptides containing β-amino acid and several nonproteogenic amino acids. It is a hepatotoxins and their primary targets are eukaryotic protein phosphatases. These toxins, produced by freshwater cyanobacteria, belong to genus *Microcystis*, *Planktothrix*, or *Anabaena* (Pearson et al. 2010). More than 90 variants of this metabolite have been identified with variable amino acids at the different positions and varying degrees of methylation (Welker and Döhren 2006). The 55 kb *mcy* cluster of *Microcystis aeruginosa* comprises 10 genes (*mcyA-J*) arranged in two divergently transcribed operons (Tillett et al. 2000). The biosynthetic scheme for microcystins shows that it starts with the loading of phenyl lactate on McyG followed by extension and methylation. The chain is then loaded on McyD for extension on two modules and then on McyE for extension, amino transfer, and condensation of D-glutamate (Tillett et al. 2000). Nodularins are other nonribosomal cyclic pentapeptides closely related to microcystins in terms of structure and biosynthesis. The biosynthetic gene cluster, *nda* has been identified in *Nodularia spumigena* and it shares strong similarities with the *mcy* gene cluster (Moffitt and Neilan 2004). Anatoxin-a and homoanatoxins-a are two potent neurotoxins produced by freshwater cyanobacteria genus *Anabaena*, *Oscillatoria*, *Phormodium*, and *Cylindrospermum*. These alkaloids act as a potent agonist of the muscle nicotinic acetylcholine receptor provoking muscle paralysis and respiratory failure. Cadel-Six et al. 2009 identified a 1.7 kb sequence called *ks2*, that produces anatoxins-*a* in *Oscillatoria* sp. The cluster designated *ana*, was then proposed to be responsible for the biosynthesis of anatoxins and a complete biosynthetic scheme for these alkaloids was proposed based on a combination of bioinformatics analysis, feeding experiments, and *in vitro* biochemical experiments on isolated enzymes (Mejean et al. 2009). Similarly, the regulation of the production of cylindrospermopsin at the transcriptional level as well as the metabolic level has been studied by Kaplan and coworkers (Shalev-Malul et al. 2008) . They concluded *Cyr* gene cluster is responsible for the cylindrospermopsin production in *Aphanizomenon* sp. The biosynthesis start with the formation of guanidinoacetate catalyzed by the amidinotransferase, CyrA. Then guanidinoacetate is thought to be loaded on the PKS CyrB for extension. Four further PKSs (CyrB, C, D, and E) are thought to extend the chain that finally leads to the desolated 7-deoxycylindrospermopsin. The other metabolites of cyanobacteria for which the biosynthetic gene clusters are now known are aeruginosins, aeruginosides, and microginins by *Microcystis* and *Planktothrix* sp.; cyanopeptolins by *Microcystis* and *Anabaena* sp.; anabaenopeptins by *Anabaena* sp.

In comparison to the fresh water form marine cyanobacteria produce a variety of secondary metabolites with interesting chemical structures and biological activities.

However, their biosynthesis remains largely unknown. Apratoxins A-G are cyclic lipopeptide from *Lyngbya bouillonii* having cytotoxic bioactivities. The biosynthetic gene clusters for apratoxin A has recently been identified by Gerwick and coworkers (Grindberg et al. 2011). They used a single cell as a PCR template to amplify the whole genome of the producer. The genome was then sequenced and the cluster was then identified using diverse screenings. The 58-kb *apr* cluster contains 12 genes among which 8 are coding for PKS or NRPs or hybrids. Curacin A is another metabolites from *Lyngbya majuscula* is of considerable interest because of its antiproliferative and antitubulinin activity and the analogues can be used as anticancer drugs. Recently, its biosynthesis has been elucidated by feeding experiments to predict the enzyme necessary to produce the compound. A 64-kb *cur* cluster was identified consisting of 14 genes, including several PKSs (Gehret et al. 2011). Jamaicamides A-C is another sodium channel blocker metabolites isolated from *Lyngbya majuscula*. The biosynthesis of this lipopeptides were studied by stable isotope incorporation and the *jam* gene cluster was identified using a combination of PCR amplification and southern hybridizations (Edwards et al. 2004) .

Cyanobacteria also produce many ribosomal peptides such as cyanobactins, the microviridins, the bacteriocins, and lantipeptides. Since the peptide core of the RPs is derived from a precursor, the clusters of genes responsible of the biosynthesis of these RPs are usually small with only a few genes coding for the modifying enzymes. These peptides show very diverse biological activities such as inhibition of proteases or cytotoxicity. However, the biosynthesis of these peptides are not fully understood, but the main steps have been deciphered (McIntosh and Schmidt 2010). The other peptide microviridins produced by different genus of cyanobacteria such as *Microcystis*, *Oscillatoria*, and *Planktothrix*. These peptides bear a unique tricyclic structure with two lactone bonds and a lactam bond and shows inhibition of protease activity. The microviridin biosynthetic gene *clusters*, *mdn*, and *mvd* have been identified and sequenced in several strains. The *mdn* cluster from *M. aeruginosa* contains five genes and on that basis the biosynthesis of microviridin B have been proposed (Ziemert et al. 2010)

Cyanobacteria produce two classes of secondary metabolites to protect them from UV radiation. In many cyanobacteria scytonemin biosynthesis occurs with exposure to UV-A and mycosporine-like amino acids with response to UV-B. The cluster of genes responsible for the production of scytonemin was identified in *Nostoc punctiforme* by random genetic inactivation. The cluster comprises six genes *Scy A-F*, directly involved in the biosynthesis of scytonemin and several other genes likely involved in the formation of precursors. The biosynthesis of this metabolite *in vivo* using isolated enzymes, *ScyA*, *B*, and *C* shows scytonemin produced from tryptophan and prephenate (Balskus and Walsh 2009). Among the 20 mycosporine-like amino acids in cyanobacteria, shinorine is well studied and its biosynthesis has been studied in *Anabaena variabilis* and *Nostoc punctiforme*. The entire gene cluster is of 6.5 kb was expressed in *E. coli* and the transformant produced shinorine. The pathway was supported by *in vitro* experiments using isolated enzymes (Balskus and Walsh 2010). The biosynthesis start from sedoheptulose-7-phosphate, which is transformed into 4-deoxygadusol. This intermediate is then condensed with glycine to give mycosporine-glycine and finally the NRPS-like protein condenses mycosporine-glycine and serene to shinorine. In the past decade, a number of novel biosynthetic gene clusters have come up due to the increasing number of availability of cyanobacterial genome sequence. The biosynthetic pathways that have been elucidated are very diverse, reflecting the chemical diversity of the secondary metabolites of cyanobacteria. However, there is a real need for more predictive bioinformatics to aid the annotation of these genes involved in secondary metabolism

References

Agger SA, Lopez-Gallego F, Hoye TR, Schmidt-Dannert C (2008) Identification of sesquiterpene synthases from *Nostoc punctiforme* PCC 73102 and *Nostoc* sp. strain PCC 7120. J Bacteriol 190:6084

Alder A, Bigler P, Werck-Reichhart D, Al-Babili S (2009) In vitro characterization of *Synechocystis* CYP120A1 revealed the first non-animal retinoic acid hydroxylase. FEBS J 276:5416–5431

Bachmann BO, Ravel J (2009) In silico prediction of microbial secondary metabolic pathways from DNA sequence data. Methods Enzymol 458:181–217

Balskus EP, Walsh CT (2008) Investigating the initial steps in the biosynthesis of cyanobacterial sunscreen scytonemin. J Am Chem Soc 130:15260–15261

Balskus EP, Walsh CT (2009) An enzymatic cyclopentyl [b] indole formation involved in scytonemin biosynthesis. J Am Chem Soc 131:14648–14649

Balskus EP, Walsh CT (2010) The genetic and molecular basis for sunscreen biosynthesis in cyanobacteria. Sci 329:1653–1656

Bernhardt R (2006) Cytochromes P450 as versatile biocatalysts. J Biotcchnol 124:128–145

Bok JW, Hoffmeister D, Maggio-Hall LA, Murillo R, Glasner JD, Keller NP (2006) Genomic mining for Aspergillus natural products. Chem Biol 13:31–37

Bok JW, Chiang YM, Szewczyk E, Reyes-Dominguez Y, Davidson AD, Sanchez JF, Lo HC, Watanabe K, Strauss J, Oakley BR, Wang CC, Keller NP (2009) Chromatin-level regulation of biosynthetic gene clusters. Nat Chem Biol 5:462–464

Cadel-Six S, Iteman I, Peyraud-Thomas C, Mann S, Ploux O, Méjean A (2009) Identification of a polyketide synthase coding sequence specific for anatoxin-a-producing Oscillatoria cyanobacteria. Appl environ microbiol 75:4909–4912

Campbell CD, Vederas JC (2010) Biosynthesis of lovastatin and related metabolites formed by fungal iterative PKS enzymes. Biopolymers 93:755–763

Cane DE (2010) Programming of erythromycin biosynthesis by a modular polyketide synthase. J Biol Chem 285:27517–27523

Challis GL (2008) Mining microbial genomes for new natural products and biosynthetic pathways. Microbiol 154:1555–1569

Chang Z, Flatt P, Gerwick WH, Nguyen VA, Willis CL, Sherman DH (2002) Genes encoding synthetases of cyclic depsipeptides, anabaenopeptilides, in *Anabaena* strain 90. Gene 296:235–247

Chughtai K, Heeren RM (2010) Mass spectrometric imaging for biomedical tissue analysis. Chem Rev 110:3237–3277

Cichewicz RH (2010) Epigenome manipulation as a pathway to new natural product scaffolds and their congeners. Nat Prod Rep 27:11–22

Comess KM, Schurdak ME (2004) Affinity-based screening techniques for enhancing lead discovery. Curr Opin Drug Discov Dev 7:411–416

Cornett DS, Reyzer ML, Chaurand P, Caprioli RM (2007) MALDI imaging mass spectrometry: molecular snapshots of biochemical systems. Nat Methods 4:828–833

Cortina NS, Krug D, Plaza A, Revermann O, Müller R (2012) Myxoprincomide: a natural product from *Myxococcus xanthus* discovered by comprehensive analysis of the secondary metabolome. Angew Chem Int Ed 51:811–816

Davies J, Ryan KS (2012) Introducing the parvome: bioactive compounds in the microbial world. ACS Chem Biol 7:252–259

DiMasi JA, Hansen RW, Grabowsk HG (2003) The price of innovation: new estimates of drug development costs. J Health Econ 22:151–185

Edwards PA, Ericsson J (1999) Sterols and isoprenoids: signaling molecules derived from the cholesterol biosynthetic pathway. Annu Rev Biochem 68:157–185

Edwards DJ, Marquez BL, Nogle LM, McPhail K, Goeger DE, Roberts MA, Gerwick WH (2004) Structure and biosynthesis of the jamaicamides, new mixed polyketide-peptide neurotoxins from the marine cyanobacterium *Lyngbya majuscula*. Chem Bio 11:817–833

Ekins S, Mestres J, Testa B (2007) In silico pharmacology of drug discovery: methods of virtual ligand screening and profiling. Br J Pharmacol 152:9–20

Erhard M, von Dohren H, Junblut P (1997) Rapid typing and elucidation of new secondary metabolites of intact cyanobacteria using MALDI-TOF mass spectrometry. Nat Biotechnol 15:906–909

Esquenazi E, Coates C, Simmons L, Gonzalez D, Gerwick WH, Dorrestein PC (2008) Visualizing the spatial distribution of secondary metabolites produced by marine cyanobacteria and sponges via MALDI-TOF imaging. Mol Biosyst 4:562–570

Exarchou V, Krucker M, van Beek TA, Vervoort J, Gerothanassis IP, Albert K (2005) LC-NMR coupling technology: recent advancements and applications in natural products analysis. Magn Reson Chem 43:681–687

Feng Z, Kim JH, Brady SF (2010) Fluostatins produced by the heterologous expression of a TAR reassembled environmental DNA derived type II PKS gene cluster. J Am Chem Soc 132:11902–11903

Gehret JJ, Gu L, Gerwick WH, Wipf P, Sherman DH, Smith JL (2011) Terminal alkene formation by the tioesterase of curacin A biosynthesis: structure of a decarboxylating thioesterase. J Biol Chem 286:14445–14454

Gessel MM, Norris JL, Caprioli RM (2014) MALDI imaging mass spectrometry: spatial molecular analysis to enable a new age of discovery. J Proteomics 107:71–82. http://dx.doi.org/10.1016/j.jprot.2014.03.021

Gonzalez DJ, Xu Y, Yang YL, Esquenazi E, Liu WT, Edlund A, Duong T, Du L, Molnar I, Gerwick WH, Jensen PR, Fischbach M, Liaw CC, Straight P, Nizet V, Dorrestein PC (2012) Observing the invisible through imaging mass spectrometry, a window into the metabolic exchange patterns of microbes. J Prot 75:5069–5076

Grata E, Boccard J, Guillarme D, Glauser G, Carrupt PA, Framer EE, wolfender JL, Rudaz S (2008) UPLC-TOF-MS for plant metabolomics: a sequential approach for wound marker analysis in *Arabidopsis thaliana*. J Chromatogr B 871:261–270

Grindberg RV, Ishoey T, Brinza D, Esquenazi E, Coates RC, Liu WT, Gerwick L, Dorrestein PC, Pevzner P, Lasken R, Gerwick WH (2011) Single cell genome amplification accelerates identification of the apratoxin biosynthetic pathway from a complex microbial assemblage. PLoS ONE 6:e18565

Gross H (2009) Genomic mining-a concept for the discovery of new bioactive natural products. Curr Opin Drug Discov Devel 12:207–219

Grünewald J, Marahiel MA (2006) Chemoenzymatic and template-directed synthesis of bioactive macrocyclic peptides. Microbiol Mol Biol Rev 70:121–146

Guiochon G, Tarafder A (2011) Fundamental challenges and opportunities for preparative supercritical fluid chromatography. J Chromatogr A 1218:1037–1114

Hasan RH, Sorkhoh NA, Bader D, Radwan SS (1994) Utilization of hydrocarbons by cyanobacteria from microbial mats on oily coasts of the Gulf. Appl Microbiol Biotechnol 41:615–619

Herrero A, Flores E (2008) The cyanobacteria. Horizon Scientific, Norfolk, pp 484

Hertweck C (2009) Hidden biosynthetic treasures brought to light. Nat Chem Biol 5:450–452

Hess WR (2011) Cyanobacterial genomics for ecology and biotechnology. Curr Opin Microbiol 14:608–614

Hoffmann D, Hevel JM, Moore RE, Moore BS (2003) Sequence analysis and biochemical characterization of the nostopeptolide A biosynthetic gene cluster from *Nostoc* sp. GSV224. Gene 311:171–180

Hopkins AL, Groom CR (2002) The druggable genome. Nat Rev Drug Discov 1:727–730

Jenke-Kodama H, Sandmann A, Müller R, Dittmann E (2005) Evolutionary implications of bacterial polyketide synthases. Mol Biol Evol 22:2027–2039

Ke N, Baudry J, Makris TM, Schuler MA, Sligar SG (2005) A retinoic acid binding cytochrome P450: CYP120A1 from *Synechocystis* sp. PCC 6803. Arch Biochem Biophys 436:110–120

Kramer R, Cohen D (2004) Functional genomics to new drug targets. Nat Rev Drug Discov 3:965–972

Kroemer RT (2007) Structure-based drug design: docking and scoring. Curr Protein Pept Sci 8:312–328

Kühnel K, Ke N, Cryle MJ, Sligar SG, Schuler MA, Schlichting I (2008) Crystal structures of substrate-free and retinoic acid-bound cyanobacterial cytochrome P450 CYP120A1. Biochemistry 47:6552–6559

Kuhnert N, Jaiswal R, Eravuchira P, El-Abassy RM, von der KB, Materny A (2011) Scope and limitations of principal component analysis of high resolution LC-TOF-MS data: the analysis of the chlorogenic acid fraction in green coffee beans as a case study. Anal Methods 3:144–155

Lee I, Oh JH, Shwab EK, Dagenais TR, Andes D, Keller NP (2009) HdaA, a class 2 histone deacetylase of Aspergillus fumigatus, affects germination and secondary metabolite production. Fungal Genet Biol 46:782–790

Liu WT, Yang YL, Xu Y, Lamsa A, Haste NM, Yang JY, Ng J, Gonzalez D, Ellermeier CD, Straight PD, Pevzner PA, Pogliano J, Nizet V, Pogliano K, Dorrestein PC (2010) Imaging mass spectrometry of intra species metabolic exchange revealed the cannibalistic factors of Bacillus subtilis. Proc Natl Acad Sci U S A 107:16286–16290

McIntosh JA, Donia MS, Schmidt EW (2010) Insights into heterocyclization from two highly similar enzymes. J Am Chem Soc 132:4089–4091

Mejean A, Mann S, Maldiney T, Vassiliadis G, Lequin O, Ploux O (2009) Evidence that biosynthesis of the neurotoxic alkaloids anatoxin-a and homoanatoxin-a in the cyanobacterium *Oscillatoria* PCC 6506 occurs on a modular polyketide synthase initiated by L-proline. J Am Chem Soc 131:7512

Moffitt MC, Neilan BA (2004) Characterization of the nodularin synthetase gene cluster and proposed theory of the evolution of cyanobacterial hepatotoxins. Appl Environ Microbiol 70:6353–6362

Molinski TF (2010) NMR of natural products at the 'nanomole-scale'. Nat Prod Rep 27:321–329

Nebert DW, Nelson DR, Feyereisen R (1989) Evolution of the cytochrome P450 genes. Xenobiotica 19:1149–1160

Oman TJ, van der Donk WA (2010) Follow the leader: the use of leader peptides to guide natural product biosynthesis. Nat Chem Biol 6:9–18

Onyango P (2004) The role of emerging genomics and proteomics technologies in cancer drug target discovery. Curr Cancer Drug Targets 4:111–124

Pearson L, Mihali T, Moffitt M, Kellmann R, Neilan B (2010) On the chemistry, toxicology and genetics of the cyanobacterial toxins, microcystin, nodularin, saxitoxin and cylindrospermopsin. Marine Drugs 8:1650–1680

Philmus B, Guerrette JP, Hemscheidt TK (2009) Substrate specificity and scope of MvdD, a GRASP-like ligase from the microviridin biosynthetic gene cluster. ACS Chem Biol 4:429–434

Piel J (2010) Biosynthesis of lovastatin and related metabolites formed by fungal iterative PKS enzymes. Nat Prod Rep 27:996–1047

Rajamani R, Good AC (2007) Ranking poses in structure-based lead discovery and optimization: current trends in scoring function development'. Curr Opin Drug Discov Devel 10:308–315

Rouhiainen L, Paulin L, Suomalainen S, Hyytiainen H, Buikema W, Haselkorn R, Sivonen K (2000) Genes encoding synthetases of cyclic depsipeptides, anabaenopeptilides, in *Anabaena* strain 90. Mol Microbiol 37:156–167

Sangeetha M, Menakha M, Vijayakumar S (2014) Cryptophycin F-A potential cyanobacterial drug for breast cancer Biomed Aging Path. 4:229–234

Schley C, Altmeyer MO, Swart R, M'uller R, Huber CG (2006) Proteome analysis of *Myxococcus xanthus* by off-line two-dimensional chromatographic separation using monolithic poly-(styrene-divinylbenzene) columns combined with ion-trap tandem mass spectrometry. J Proteome Res 5:2760–2768

Schmidt EW, Nelson JT, Rasko DA, Sudek S, Eisen JA, Haygood MG, Ravel J (2005) Patellamide A and C biosynthesis by a microcin-like pathway in *Prochloron didemni*, the cyanobacterial symbiont of *Lissoclinum patella*. Proc Natl Acad Sci U S A 102:7315–7320

Schmitt EK, Moore CM, Krastel P, Petersen F (2011). Natural products as catalysts for innovation: a pharmaceutical industry perspective. Curr Opin Chem Biol 15:497–504 (Excerpts used with permission)

Schroder FC, Farmer JJ, Attygalle AB, Smedley SR, Eisner T, Meinwald J (1998) Combinatorial chemistry in insects: a library of defensive macrocyclic polyamines. Sci 281:428–431

Schwarzer D, Finking R, Marahiel MA (2003) Nonribosomal peptides: from genes to products. Nat Prod Rep 20:275–287

Seeley EH, Caprioli RM (2008) Molecular imaging of proteins in tissues by mass spectrometry. Proc Natl Acad Sci U S A 105:18126–18131

Shalev- Malul G, Lieman- Hurwitz J, Viner- Mozzini Y, Sukenik A, Gaathon A, Lebendiker M, Kaplan A (2008) An AbrB- like protein might be involved in the regulation of cylindrospermopsin production by *Aphanizomenon ovalisporum*. Environ Microbiol 10:988–999

Shanta SR, Zhou LH, Park YS, Kim YH, Kim Y, Kim KP (2012) Binary matrix for MALDI imaging mass spectrometry of phospholipids in both ion modes. Anal Chem 83:1252–1259

Shastri AA, Morgan JA (2005) Flux balance analysis of photoautotrophic metabolism. Biotechnol Prog 21:1617–1626

Singh S, Verma SK (2012) Application of direct analysis in real time mass spectrometry (DART-MS) for identification of an epiphytic cyanobacterium, *Nostoc* sp. Anal Lett 45:2562–2568

Soule T, Stout V, Swingley WD, Meeks JC, Garcia-Pichel F (2007) Molecular genetics and genomic analysis of scytonemin biosynthesis in *Nostoc punctiforme* ATCC 29133. J Bacteriol 189:4465–4472

Staunton J, Weissman K (2001) Polyketide biosynthesis: a millennium review. J Nat Prod Rep 18:380–416

Szewczyk E, Chiang YM, Oakley CE, Davidson AD, Wang CC, Oakley BR (2008) Identification and characterization of the asperthecin gene cluster of *Aspergillus nidulans*. Appl Environ Microbiol 74:7607–7612

Thomason L, Court DL, Bubunenko M, Costantino N, Wilson H, Datta S, Oppenheim A (2007) Recombineering: genetic engineering in bacteria using homologous recombination. Curr Protoc Mol Biol. doi:10.1002/0471142727.mb0116s106 (Chap. 1: Unit 1.16)

Tillett D, Dittmann E, Erhard M, von Döhren H, Börner T, Neilan BA (2000) Structural organization of microcystin biosynthesis in *Microcystis aeruginosa* PCC7806: an integrated peptide-polyketide synthetase system. Chem Biol 7:753–764

Torres S, Fjetland CR, Lammers PJ (2005) Alkane-induced expression, substrate binding profile, and immunolocalization of a cytochrome P450 encoded on the nifD excision element of Anabaena 7120. BMC Microbiol 5:1–12

Velasquez JE, van der Donk WA (2011) Genome mining for ribosomal synthesized natural products. Curr Opin Chem Biol 15:11–21

Vijayakumar S, Menakha M (2013) Tasiamide-B a new cyanobacterial compound for treating skin cancer. Biomed Prev Nutr 4:355–358. http://dx.doi.org/10.1016/j.bionut.2013.10.001

Vijayakumar S, Menakha M (2014) Prediction of new cyanobacterial drug for treating lung cancer. Biomed Aging Patho 4:49–52

Walsh CT, Fischbach MA (2010) Natural products version 2.0: connecting genes to molecules. J Am Chem Soc 132:2469–2493

Weber T (2014) In silico tools for the analysis of antibiotic biosynthetic pathways. Int J of Med Microbiol 304:230-235 (Excerpt used with permission)

Welker M, Von Döhren H (2006) Cyanobacterial peptides-nature's own combinatorial biosynthesis. FEMS Microbiol Rev 30:530–563

Welker M, Marsalek B, Sejnohova L, von Dohren H (2006) Detection and identification of oligopeptides in *Microcystis* (cyanobacteria) colonies: toward an understanding of metabolic diversity. Peptides 27:2090–2103

Wolfender JL, Marti G, Queiroz EF (2010) Advances in techniques for profiling crude extracts and for the rapid identification of natural products: dereplication, quality control and metabolomics. Curr Org Chem 14:1808–1832

Xu Y, Alvey RM, Byrne PO, Graham JE, Shen G, Bryant DA (2011) Expression of genes in cyanobacteria: adaptation of endogenous plasmids as platforms for high-level gene expression in Synechococcus sp. PCC 7002. Methods Mol Biol 684:273–293

Yang YL, Xu YQ, Straight P, Dorrestein PC (2009) Translating metabolic exchange with imaging mass spectrometry. Nat Chem Biol 5:885–887

Ziemert N, Ishida K, Liaimer A, Hertweck C, Dittmann E (2008) Ribosomal synthesis of tricyclic depsipeptides in bloom-forming cyanobacteria. Angew Chem Int Ed Engl 47:7756–7759

Ziemert N, Ishida K, Weiz A, Hertweck C, Dittmann E (2010) Exploiting the natural diversity of microviridin gene clusters for discovery of novel tricyclic depsipeptides. Appl Env Microbiol 76:3568–3574

Chapter 3
Glycoconjugates of Cyanobacteria and Potential Drug Development From Them

3.1 Glycoconjugates of Cyanobacteria

Carbohydrates, in addition to acting as an energy source, play diverse roles in the life cycle. Complex carbohydrates are the next frontier in understanding the secret molecular messages that rule the life of our cells. Carbohydrates determine blood type, regulate plant growth, and have roles in cancer, diabetes, and human development. It is the major component in cell-wall composition and responsible for the integrity of microbial pathogens (Moran et al. 2005). Glycomics (the study of carbohydrate-related processes, including carbohydrate metabolism, glycoconjugates structures, and cell–cell interactions) has now become a well-established area of study and microbial glycomics provides desperately needed new drug targets. So glycobiology is an extremely broad field, encompassing the study of any molecule that contains a carbohydrate as part of its structure. Glycoconjugates are formed when mono-, oligo-, or polysaccharides are attached to proteins or lipids. The sugar-containing portions of the resulting glycoproteins and glycolipids are generally complex heteropolymers and referred to as glycans. The majority of glycans are either N-linked or O-linked depending on the attachment to the proteins or lipids through a nitrogen atom or through an oxygen atom. In addition to DNA and proteins, glycans represent the third dimension in molecular biology. Glycan structures are encoded indirectly in the genome. The sugar structures are not encoded directly in the DNA sequences, but are determined by transcription and translation of genes to generate glycosyltransferases that in turn control the synthesis of the glycan portions of glycoconjugates. Glycobiology has become the subject of much research during the past few years and holds a wealth of hope for the future medicine. In addition, it is also an innovative scientific field that may help us to better understand and combat the visible signs of aging and reduce the effects of time on the skin.

In natural habitat, many cyanobacteria form visible colonies that consist of biochemically complex extracellular matrices and cellular filaments embedded within extracellular polysaccharides (EPS), accounting for 60–80 % of the dry mass (Hill et al.1997). The production of EPS in cyanobacteria is widely known (Gloaguen

© Springer International Publishing Switzerland 2015
S. Mandal, J. Rath, *Extremophilic Cyanobacteria For Novel Drug Development*,
SpringerBriefs in Pharmaceutical Science & Drug Development,
DOI 10.1007/978-3-319-12009-6_3

et al. 1995). Many cyanobacteria show little to no metabolic activity and are able to rapidly resume metabolism upon rehydration—a phenomenon termed "anhydrobiosis" (Crowe 2002). The terrestrial cyanobacterium *Nostoc commune* can retain viability for over 100 years upon desiccation (Lipman 1941; Cameron 1962), and the complex glycan play an important role in the survival of these organisms under stress conditions. The glycans of cyanobacteria remain associated with the cell surface as sheaths, capsules and/or slimes, or be liberated into the surrounding environment as released polysaccharides (RPS; Pereira et al. 2009). The RPS serves as a boundary between the cyanobacterial cell and its immediate environment. Many putative roles have been proposed for these polymers, such as protection against dehydration or UV radiation, biomineralization, and phagocytosis. It has also been postulated that RPS confer on the cyanobacteria the ability to adhere to solid substrates, as well as facilitate the locomotion of gliding strains.

Most cyanobacterial glycans are characterized by the presence of uronic acids, pentoses, a polypeptide moiety, or other nonsaccharide components, such as organic (e.g., acetyl, pyruvyl, succinyl group) or inorganic (e.g., sulfate or phosphate group) substituents (De Philippis and Vincenzini 1998). Therefore, cyanobacteria are regarded as a very abundant source of structurally diverse glycans, a feature rarely found in the polymers released by strains belonging to other microbial groups (Sutherland 1994). Cyanobacterial released polysaccharide often shows the presence of one or two pentose sugar that are usually absent in other polysaccharide of prokaryotic origin (Sutherland 1994). This moiety protects the neighboring glycosidic bonds from the more common glycan hydrolases. Most RPS synthesized by cyanobacteria are quite complex, being composed of six or more monosaccharides. This is the striking difference from the polymers synthesized by other bacteria or by microalgae, in which the number of monomers is usually less than four. Usually, glucose is the dominant monosaccharide of the RPS although ironic acid, xylose, arabinose, fucose, rhamnose, and mannose are the dominant monosaccharides in some cyanobacterial RPSs. Galactose is the dominant sugar only in *Anabaena sphaerica*. Ribose, osamine, methyl sugar, and unidentified residues are also present in several cyanobacterial RPSs. The polysaccharides produced by *N. commune*, *N. insulare*, and *Cyanothece* sp. are composed of repeating units of six, four, and three monosaccharides, respectively (Volk et al. 2007). On the other hand, the structures proposed for the EPS produced by *Mastigocladus laminosus* and *Cyanospira capsulata* are far more complex, with repeating units of fifteen and eight monosaccharides, respectively (Gloaguen et al. 1999). For *Spirulina platensis*, no structure was proposed, but it was demonstrated that its RPS repeating unit contains at least 15 sugar residues (Filali Mouhim et al. 1993). The high number of different monosaccharides found in cyanobacterial glycan and the consequential variety of linkage types is usually considered a reason for the presence of complex repeating units, as well as for a broad range of possible structures and architectures of these macromolecules. A large number of different monosaccharides in only one polymer make many structures and architectures possible (Atkins 1986), thus increasing the chance of having a polymer with peculiar properties, not common to currently utilized products. The rheological properties of RPSs' aqueous solutions make them useful as thickening agents for water solutions, together with their ability to stabilize the flow properties of their own solutions under drastic changes of pH, temperature, and ionic strength (Sutherland 1998).

The released polysaccharides in many cyanobacteria attached to proteins and characterized by the presence of pyruvate and acetate groups. Polypeptides enriched with glycine, alanine, valine, leucine, isoleucine, and phenylalanine have been reported in the RPS of *C. capsulate* (Marra et al. 1990). A most interesting feature is the presence of sulfate groups in the RPSs and CPSs of many strains of cyanobacteria. Differences in the sulfur content and in the monosaccharidic composition reported for the sheath and RPS of some cyanobacteria strongly support the hypothesis of their origin from different biosynthetic pathways (Micheletti et al. 2008). The sulfate groups and the uronic acids contribute to the anionic nature of the RPS, conferring a negative charge and a "sticky" behavior to the overall macromolecule (Mancuso Nichols et al. 2005). All these structural factors, i.e., the large numbers of constitutive monosaccharides, the variety of their linkages in the polymeric chains, the variability of functional groups, as well as the protein content give rise to very complex structures of cyanobacterial glycans and impotence in glycomics.

3.2 Glycan Biosynthesis in Cyanobacteria

Glycan biosynthetic pathways are very complex, including, the enzymes directly involved in the polysaccharide synthesis, enzymes engaged in the formation of the cell-wall polysaccharides and lipopolysaccharides (Mozzi et al. 2003). However, cyanobacteria have not been thoroughly examined and, consequently, the information available is extremely limited (Yoshimura et al. 2007). Typically, the biosynthesis of glycans comprises four distinct steps occurring in different cellular compartments. First, the activation of the monosaccharides and conversion into sugar nucleotides in the cytoplasm, then the assembly of the repeating units by sequential addition of sugars onto a lipid carrier by glycosyltransferases at the plasma membrane, followed by the polymerization of the repeating units in the periplasmic face of the plasma membrane and finally the export of the polymer to the cell surface (Sutherland 2001). Among the genes involved are rfbABCD, also frequently called rml genes (Reeves et al. 1996), which encode proteins involved in the biosynthesis of L-rhamnose. Rhamnose is one of the sugars frequently found in the cyanobacterial RPS, and the proteins encoded by the rfb genes are listed in CyanoBase (http://bacteria.kazusa.or.jp/cyanobase/) as involved in the assembly of cyanobacterial surface polysaccharides. However, the presence of several acidic or neutral monosaccharides in cyanobacterial glycan indicates that their biosynthetic pathway may be even more complex (Li et al. 2002). The first insight into the genes encoding proteins involved in EPS assembly and export in cyanobacteria was recently done by Pereira et al. 2009. They performed *in silico* analysis of available cyanobacteria genome sequences, revealing the existence of genes coding for proteins that possess the conserved domains involved in bacterial EPS assembly and export. The results also show that in cyanobacteria, the EPS-related genes often occur in multiple copies scattered throughout the genomes, either isolated or in small clusters (Pereira et al. 2009). Glycosyltransferases are key enzymes for the biosynthesis of the RPS repeating unit, catalyzing the transfer of the sugar nucleotides of activated donor molecules to specific receptor molecule. Despite the increasing interest in these polymers, the information about their biosynthetic pathways is still limited. Studies performed in other bacteria revealed that the mechanism of EPS assembly and export are relatively conserved, generally following the Wzy-dependent or the ABC-dependent pathways, which require the involvement of polysaccharide copolymers (PCP) and outer membrane polysaccharide export (OPX) proteins. In cyanobacteria, the genes encoding these proteins occur in multiple copies, scattered throughout the genome, either isolated or in small clusters. However, it is necessary to identify other genes that may be related to this process, understand their genomic distribution, and reconstruct their evolutionary history. *In silico* analysis of the cyanobacterial genomes revealed the presence of numerous genes putatively encoding glycosyltransferases. However, the enzymes have not been biochemically characterized, which makes it impossible to assign their function to the synthesis of RPS. The synthesis and secretion of RPS in cyanobacteria probably follow the same pathways as with other bacteria. However, the presence of high numbers of different sugars in cyanobacterial RPS suggests that the synthesis of the sugar nucleotides is more complex, involving a higher number of different enzymatic reactions.

3.3   Glycotherapy Using Cyanobacterial Glycoconjugates

Glycans play many important roles in biological processes, including involvement in the processes of cellular trafficking, cellular differentiation, cell–cell adhesion, hormone–cell recognition, and viral-host cell and bacterial-host cell attachment. Glycans and glycoconjugates are linked to a number of serious medical conditions such as malignancies, Creutzfeldt–Jakob disease, cystic fibrosis, and inflammatory diseases, cardiovascular diseases, microbial infections, and lysosomal storage diseases (Axford 2001). Glycans also have important constituents of medicinal important agents, such as the antitumor antibiotics, cardiac glycosides, and gangliosides. The oligosaccharide moieties of the antitumor antibiotics ciclamycin and calicheamicin are important in the DNA recognition critical for the activity of the compound. Cardiac glycosides, such as digoxin are steroids conjugated to a trisaccharide. Gangliosides are lipid-olegosaccharide conjugates, and have been used to treat spinal cord injury. The use of saccharides as rigid scaffolding for presenting key functional elements in a defined arrangement for the identification of new receptor ligands is another area of great promise. Saccharide–receptor interactions are implicated in many disease processes, including infections, inflammatory responses, and cancer metastasis. Approaches designed to disrupt such interactions are relevant to the treatment of these disorders. Rare genetic disorders of carbohydrate synthesis or metabolism may in some cases be treated with carbohydrate moieties or by enzyme replacement therapy designed to restore the deficient pathways.

Sulfated homopolysaccharides and heteropolysaccharides isolated from a number of cyanobacteria have demonstrated potent activity against retroviruses, including human immunodeficiency virus (HIV) and herpes simplex virus (Schaeffer and Krylov 2000). The polysaccharides inhibit the viral cytopathic effect and also prevent HIV-induced syncytium formation. Lectins are carbohydrate-binding proteins that do not modify the carbohydrates to which they bind. Several lectins from cyanobacteria were shown to be able to bind to HIV-1 in a carbohydrate-dependent manner. The cyanobacterial lectins Cyanovirin-N (CV-N), Microvirin (MVN), Scytovirin (SVN), and *Oscillatoria agardhii* agglutinin (OAA) are promising candidate microbicides for the prevention of HIV transmission by interacting with the glycans on HIV gp120. Cyanovirin-N (CV-N), derived from the cyanobacterium *N. ellipsosporum*, has been given by far the most attention as antiviral lectin. The predominant form of CV-N in solution is the monomeric form, and CV-N contains two carbohydrate recognition sites on symmetrically opposed domains of the protein, so it can cross-link branched oligomannosides to form higher order structures (Bewley et al. 1998). Cyanovirin-N was active against HIV-1 and HIV-2 laboratory strains with EC_{50} values between 0.1 and 160 nM. CV-N was also tested against clinical isolates from many different clades and was found active with EC_{50} values between 0.3 and 160 nM and was found active against Simian Immunodeficiency Virus (SIV; Xiong et al. 2010). Microvirin (MVN) has been a recently discovered novel antiviral lectin isolated from the cyanobacterium *Mycrocystis aeruginosa* (Huskens et al. 2010). *M. viridis* lectin (MVL) another antiviral lectin isolated from the cyanobacterium *Mycrocystis viridis* (Bewley et al. 2004). Scytovirin (SVN) isolated from aqueous extracts of the cultured cyanobacterium *Scytonema varium* binds to a specific tetrasaccharide substructure of the high-mannose oligosaccharide. SVN was active against HIV-1 laboratory strains and clinical isolates with EC_{50} values between 0.3 and 7 nM and 0.4 and 394 nM, respectively (Alexandre et al. 2010). The cyanobacterial lectin *O. agardhii* agglutinin (OAA) was isolated from *Oscillatoria agardhii* and preferentially binds to Manα(1–6)Man. OAA was first described by Sato et al. 2000 was active in MT-4 cells against the X4 HIV-1 strain IIIB with an EC_{50} of 44.5 nM. These HIV-infected cells express the viral glycoproteins on their cell surface and the lectins of cyanobacteria are able to bind and inhibit syncytium formation between infected cells and uninfected CD4[+] T cells. Cyanobacterial lectins CV-N, MVN can inhibit the transmission of HIV between persistently infected cells and uninfected CD4[+] T cells with IC_{50} values of 4–46 nM, 124 nM and < 1 nM, respectively (Mori et al. 2005; Buffa et al. 2009). Besides, as potential microbicide candidate, CV-N was also tested in ex vivo cervical explant model and it could potently inhibit infection in these models (O'Keefe et al. 2009). In another preclinical test, CV-N was evaluated in a novel penile tissue explant model (Fischetti et al. 2009) and CV-N conferred 95 % protection against HIV-1at 1 μM, which is similar to that seen in cervical explants (Fischetti et al. 2009). HIV-1 infection is commonly associated with other sexual transmitted viruses and CV-N is effective against Ebola virus, influenza A and B, hepatitis C virus, measles virus, herpes simplex virus type-1 (HSV-1), and human herpes virus 6 (HHV-6) (Xiong et al. 2010; Buffa et al. 2009; Dey et al. 2000). Many studies have already highlighted that the polysaccharides released into the culture medium by cyanobacteria present antiviral bioactivity against different kinds of viruses, either mammalian or otherwise. Radonic et al. 2010 and Chen et al. 2010 have reviewed the antiviral effects on different host cell-lines. Calcium-Spirulan, an intracellular polysaccharide produced by *Arthospira platensis*, inhibited the replication of several viruses *in vitro* by inhibiting the penetration of the virus into the different host cells used (Hayashi et al. 1996). Radonic and coworkers also showed that the polysaccharide released into the medium by *A. platensis* exhibited antiviral activity *in vitro* and *in vivo* against two strains of *Vaccinia* virus and an *Ectromelia* virus. Moreover, a novel acidic polysaccharide, nostoflan, was isolated from *Nostoc flagelliforme*, and was verified to possess a remarkable anti-viral effect on a variety of enveloped viruses, including HSV-1, human cytomegalo virus, and the influenza A virus (Blinkova et al. 2001).

The formation of cancer cells in the human body can be directly induced by free radicals. Natural anticancer drugs as chemo-preventive agents have gained a positive popularity in the treatment of cancer. Hence, radical scavenging compounds such as sulfated polysaccharides (SPS) of cyanobacteria can be used to reduce cancer formation in the human body. During the cancer multistage cascade, normal cells undergo initiation, promotion, and progression processes. Most natural anticancer compounds are able to manipulate the growth of cancer cells with no or minor side effects. Therefore, identification of novel effective cancer chemo-preventive agents has become an important worldwide strategy in cancer prevention. Many studies have suggested that polysaccharides can inhibit tumor growth by prevention of tumorigenesis by oral consumption of active preparations; and direct anticancer activity, such as the induction of tumor cell apoptosis or immune potentiation activity in combination with chemotherapy; and the inhibition of tumor metastasis. One potentially promising activity of cyanobacterial polysaccharides is the ability to prevent tumour cell growth. Calcium-Spirulan of *Arhrospira* (*Spirulina*) was reported to prevent pulmonary metastasis, also preventing the adhesion and proliferation of tumor cells. It is also promising in treating spinal cord injuries and as matrices for stem-cell cultures (Morais et al. 2010). Some antimicrobial compounds have been detected in both the cellular extracts and the extracellular products of some *Nostoc* species. The hot-water extracts from *N. flagelliforme* have been reported to evidence antitumor activity, and this effect may be attributable to the polysaccharides. The presence of sulfated group has indicated that many cyanobacterial exopolysaccharides possess promising potential applications in the pharmaceutical industry. The capsular polysaccharide produced by the thermophilic cyanobacterium *Mastigocladus laminosus* inhibited the proliferation of A431 human epidermoid carcinoma cell line in a dose-dependent manner with an IC_{50} value of 50 mg mL ± 1. It has been proposed that the purified polysaccharide fraction act via the inhibition of metalloproteases (MMP) expression. Taking into account the rising trend of the incidence of cancers of various organs, effective therapies are urgently needed to control human malignancies. However, almost all of the chemotherapy drugs currently on the market cause serious side effects. Several polysaccharides and polysaccharide–protein complexes possess anticancer activities or can increase the efficacy of conventional chemotherapy drugs.

Glycans are water soluble. Hence, attaching a glycosidic moiety into a target drug molecule increases its hydrophilicity. This influences tremendously the pharmacokinetic properties of the compound, e.g., circulation, elimination, and concentrations in the body fluids. In this context, glycans of cyanobacteria are having tremendous therapeutic value. Interesting studies concern with the roles of carbohydrates as recognition sites on the cell surfaces (Ofek et al. 1978). It has been demonstrated that microorganisms must specifically attach to the host cell to avoid being washed away by secretions. Such attachment would permit colonization or infection, sometimes followed by membrane penetration and invasion. Bergey and Stinson 1988 and Bellamy et al. 1993 provided some evidence of the participation of the carbohydrates in the cellular recognition process of the interaction between host and pathogen. It is an attractive possibility to use carbohydrate-based drugs for blocking the early stages of an infection process. In particular, SPS are involved in biological activities such as cell recognition, cell adhesion, or regulation of receptor functions (Cassaro and Dietrich 1977). Obtaining SPS from cyanobacteria has made development of biotechnology for obtaining new therapeutic products easier. Ascencio et al. 1993 identified heparan sulfate glycosaminoglycan as a putative host target for *Helicobacter pylori* adhesion. A number of marine and freshwater cyanobacteria strains for the production of sulfate exopolysaccharides were screened to evaluate whether these exopolysaccharides can block adherence to human and fish cells of human gastrointestinal pathogens, such as *H. pylori* and *Aeromonas veronii*. Results indicate that SPS of some species of cyanobacteria inhibit the cytoadhesion process of *H. pylori* to animal cells. Therefore, the treatment with SPS could be used to block the initial process of colonization of the host by *H. Pylori*, so it may represent an alternative prophylactic therapy in microbial infections where the process of cytoadhesion of host to the pathogen is likely to be blocked. Such a therapy could replace the use of antibiotics and antiparasitic drugs that are always aggressive toward the host and, moreover, can generate resistant strains to the antibiotic, making it necessary to use second generation antibiotics (SjÖustrÖum et al. 1996).

3.4   Cyanobacterial Glycans as Drug Delivery Device

Glycans of cyanobacteria have a great potential as binders, fillers, coatings and slow release agents for orally administered drug preparations. Several common problems are encountered in attempts to successfully administer drugs by conventional oral or nasal routes. One major concern is the time available for drug absorption may be too short, and the drug may be degraded or mechanically swept away before being absorbed. Bioadhesive drug delivery is a novel approach in which the drug is coupled to a carrier molecule that adheres to the biomembrane of the target cell. Cyanobacterial lectins are useful in this respect as they could be attached specifically to gut or nasal mucosa cells through the recognition of glycoconjugates present at the cell surface. This adhesion could allow the drug time to cross the plasma membrane successfully. Lectins can be also used for directing drugs to the surface of bacteria responsible for human disease. The cell walls of Gram-positive bacteria are made of peptidoglycans and those of Gram-negative bacteria are lipopolysaccharides. These cell surface glycoconjugates are potential binding sites for lectins and offer a means of targeted drug delivery. Asialoglycoprotein receptor is found in mammalian hepatocytes and this has high specificity for ligands displaying terminal Gal or GalNAc residues. Liver-specific carriers, such as liposomes and polymers that display these residues, have been created for the selective delivery of drugs and genes to hepatocytes. This approach could be extended to allow more general carbohydrate-directed therapies (Duncan 2003). A novel strategy, lectin-directed enzyme-activated prodrug therapy (LEAPT), has recently been developed to allow directed delivery of drugs to specific organs that display lectins. First, a nonmammalian glycosidase enzyme (e.g., -L-rhamnosidase from *Penicillium decumbens*, naringinase) is localized to the target cell by carrier carbohydrate–lectin interactions. A prodrug that must be a substrate for the glycosidase enzyme is then administered. Since drug release will only occur within the vicinity of the glycosidase enzyme, and this is localized to the target cell, selective drug release occurs at the target cell. By altering the carbohydrate attached to the glycosidase enzyme, it is possible to target different cell types (Davis 2002).

Hydrogels are three-dimensional, hydrophilic, and polymeric networks capable of imbibing large amounts of water or biological fluids (Peppas et al. 2000). Due to their high-water content and soft consistency, hydrogels are very similar to natural living tissues, with performances that overcome the other classes of biomaterials. Among the natural polymers, cyanobacterial polysaccharides have a wide application due to their peculiarities and can act as hydrogels. They are abundant and obtained from renewable sources, present a large variety of composition and properties, not easily reproducible by synthetic routes and their production is generally easier and cheaper than for synthetic polymers. It is generally accepted that the drug release from the matrix follows two main mechanisms, i.e., the diffusion of the protein through the pores of the polymer network and degradation of the polymer network (Gombotz and Wee 1998). In addition, water diffusion and swelling through the hydrogen are two of the major factors affecting drug release rate (Faroongsarng and Sukonrat 2008). Thus, drug release from a hydrogel is governed by several parameters: pore-volume fraction, pore sizes, extent of interconnections, size and physico-chemical nature of the drug molecules, and in general, type and strength of interactions between drugs and polymeric chains (Hoffman 2002). Hydrogel-based systems have been exploited for protection and delivery of low-molecular weight drugs and macro-molecular payloads (Agnihotri et al. 2004, Oh et al. 2009), such as peptide and protein drugs (insulin, melatonin, heparin, haemoglobin, parathyroid hormone, and calcitonin, etc.; He et al. 1999), nucleic acids (DNA, siRNA; Borchard 2001), and antigens (from pathogens responsible of influenza, pertussis, diphtheria, and tetanus, etc.; Illum et al. 2001). In addition to antigens, vaccines have been encapsulated (attenuated or inactivated pathogens) in both medical and veterinary fields, combining the cell immobilization to a drug delivery system. Such technology is gaining greater attention, due to the possibility to protect the vaccine during the administration (also for the oral route) and hopefully to control the targeting and delivery rate (Año et al. 2011).

Nanotechnology plays an indispensable role in the field of advanced medicine and biotechnology. Cyanobacterial polysaccharides can be also used as natural biopolymers at the nanoscale level by biopolymeric nanocapsules, nanogels, and nano particles formulations for application to the encapsulation of biological entities of smaller dimensions (mainly drugs, therapeutic proteins, and nucleic acids; Skiba et al. 1995). Several cyanobacterial species were also studied for the extraction of highly viscous and biologically active SPS (Richert et al. 2005); *Aphanothecasacrum* is a typical example of cyanobacteria with pharmaceutically important sulfated fucose in its jelly matrix (Okajima-Kaneko et al. 2007). Polysaccharides are widely used in the drug delivery applications as polyelectrolytes forming multilayers based on various factors like hydration, internal composition, charge distribution, and chemical modification, etc. There are two major elements which make polysaccharides an important material to use in bio-nanotechnology: (1) the glycosidic bonds, which can be easily cleaved by hydrolase enzymes and hence they are biodegradable; (2) the presence of positive ammonium (NH_3^+) group or negative carboxyl (COO^-) and sulfate (SO_3^-) groups makes them behave like polyelectrolytes, which self assemble during nano or microparticle formation. Owing to the potential bio-physico-chemical properties, most of the SPS of cyanobacteria are important for their nano level applications like nanoparticle formulation and targeted delivery of drugs as well as bioactive molecules. So far, various natural and synthetic SPS were used for the preparation of microspheres and nanospheres in order to deliver various drugs as well as biologically active molecules (Crouzier et al. 2010). Most of the cyanobacterial exopolysaccharides (EPS) are highly sulfated compared to the bacterial polysaccharides, which generally lack the sulfate moiety. Hence, bacterial polysaccharides are comparatively less important than cyanobacterial polysaccharides in biomedical applications. Drug delivery systems based on cyanobacterial polysaccharide nanoparticles present several advantages such as the capability to penetrate cells and tissues; the possibility to improve the bioavailability of drugs, reducing toxic side effects; the ability to control release properties due to the biodegradability and the stimuli (pH, ion and temperature), and sensibility of materials (Liu et al. 2008). SPS are universally accepted safe and biocompatible materials employed in modern pharmaceutical research. Their nontoxic behavior and immune modulating effects have to be verified more through clinical trials for future applications. Use of natural SPS nanoparticles in medicine and pharmacotherapy can replace several chemical drugs that induce vulnerable side effects; thus bringing a cost-effective mode of treatment, thereby improving patient compliance. The rationalization of the assembly mechanisms and the capability to tailor the properties (size, charge, and loading capability) to desirable levels are essential goals to advance biodegradable polysaccharidic nanoparticles as efficient drug delivery vehicles.

3.5   Glycoconjugate Vaccines

Vaccination is the most cost efficient and powerful medical intervention in control, prevention, and eradication of many diseases that affect human populations. In the past decade, increase in the resistance to antibiotics has been observed, leading to a serious threat for successful treatment of bacterial infections. This feature in addition to difficulties in developing adequate drugs against (tropical) diseases caused by parasites has stimulated the interest in vaccines to prevent infections. Vaccination is the active immunization with an immunogen (the vaccine) administered so that the host develops specific antibodies and B- and T-memory cells that can act against the natural immunogen. These cells stand ready to be activated should the host later be exposed to the pathogen bearing the natural immunogen. Polysaccharides are considered to give an immune response independently of T cells; they stimulate B cells to produce antibodies without the involvement of T cells. However, some zwitterionic capsular polysaccharides can activate $CD4^+$ T cells. In contrast to polysaccharides, glycoproteins are T-cell dependent (TD) antigens, having a larger immune response to the same antigens. Avery and Goedel 1931 reported that covalent attachment of carbohydrates to a suitable protein as a source of T-cell epitopes induced an enhanced immunogenicity compared to the polysaccharides alone. In general, immunization with neoglycoproteins consisting of a capsule-derived carbohydrates coupled to an immunogenic protein provides a long-lasting protection to encapsulated bacteria for adults as well as for persons at high risk and young children. Many currently utilized vaccines are comprised of glycans including vaccinations against *Neisseria meningitides* (Menactra), *Streptococcus pneumonia* (Prevnar), *Haemophilus influenzae* type b (Hib; Hiberix, Comvax), and *Salmonella typhi* (Morelli et al. 2011).

For the design of modern carbohydrate-based glycoconjugate vaccines, four important considerations are generally applicable; the antigen source, the carrier, the conjugation method, and the adjuvant. Glycan antigens are diverse and range from large, elaborate capsular polysaccharides to small monosaccharide tumor antigens. In general, polysaccharides exist as a family of closely related species that vary in their degree of polymerization. As the pertinent immunogenic epitopes comprise only part of the glycan, oligosaccharides are often adequate for vaccine preparation. These molecules may be derived from digestion of naturally derived polysaccharides or produced as a chemically homogeneous species through synthetic methods. Carriers are most often proteins and could be toxoids, keyhole limpet haemocyanin (KlH), or virus capsids. They should be immunogenic and express multiple loci for conjugation, as polyvalent display is crucial for generating carbohydrate-specific antibody responses. Coupling of oligosaccharide antigens to the carrier necessitates the activation of the sugars and/or the carrier. Several procedures have been developed to activate polysaccharides, but most result in the creation of reactive groups that are randomly distributed throughout the polymer. To generate well-defined conjugates, the linkage between sugar and carrier should be as specific as possible. Several synthetic linkers are available, but one must be cautious of the immunogenicity of these linkers relative to the glycan antigen (Buskas et al. 2004; Ni et al. 2006). The immune response may be predominantly directed against the linker and away from the carbohydrate antigen. Also, steric issues may be addressed with bifunctional spacers to enhance the efficiency of loading. Finally, adjuvants are often included to improve the immunogenicity of the target carbohydrate antigens. Alum is the only adjuvant approved for human use in the USA, however, several promising formulations are in clinical trials.

The identification of potential carbohydrate epitope targets is typically the first step in the development of a carbohydrate vaccine. Analysis and purification of naturally occurring glycans are complicated by the micro-heterogeneity of carbohydrates arising from non-template-driven biosynthesis. In addition, there is a trade-off between the ability to perform high-throughput analysis of glycan mixtures and the ability to do fine structure characterization. Biochemical and analytical methods such as nuclear magnetic resonance (NMR), electrospray ionization–mass spectrometry (ESI–MS), matrix-assisted laser desorption ionization MS (MALDI–MS) and capillary electrophoresis have been developed to profile the repertoire of carbohydrate structures isolated from cells and tissues (Raman et al. 2005). Each methodology is best suited for determining a certain set of attributes; for example, chain length and mass composition relationships from MALDI–MS or monosaccharide composition and linkage information using NMR. To improve the utility of these methods and to enable high-throughput analysis, informatics-based sequencing methodologies are being used to integrate information from multiple complementary techniques (Guerrini et al. 2002, Venkataraman et al. 1999). Structural data of protein–glycan interactions may provide valuable insights into the appropriate presentation of glycan antigens. Unfortunately, these data are difficult to obtain and consequently, relatively rare. A significant challenge stems from the flexibility of glycans, which assume multiple conformational states under physiological conditions. Currently, the combination of X-ray crystallography, NMR and molecular dynamics simulations, and other computational methods (such as docking simulations and absolute and relative free-energy calculations), are used in structural glycobiology to elucidate the conformations of free carbohydrates and protein–carbohydrate complexes (Wormald et al. 2002; Jimenez-Barbero et al. 1999). Emerging technologies include oxidative foot-printing of carbohydrate binding surfaces, which complements epitope mapping techniques, such as saturation transfer difference NMR (DeMarco and Woods 2008). Another promising approach for three-dimensional carbohydrate–protein complex determination combines partially oriented NMR spectroscopy with computational simulations (DeMarco and Woods 2008).

For several types of bacterial infections glycoconjugate vaccines can be based on fragments of capsular polysaccharides. The availability of methods to prepare specific oligosaccharide structures opened the possibility to explore the relation between the oligosaccharide chain length and the potency of the glycoconjugate as a vaccine. In a study on the immunogenicity and protective capacity of *S. pneumoniae* 6B capsular polysaccharide derived di-(Rhaa1-4-Rib-ol-5P-), tri-(Rib-ol-5P-2Gala1-3Glc-) and tetrasaccharide (Rhaa1-4-Rib-ol-5P-2Gala1-3Glc-) conjugated to carrier protein keyhole limpet hemocyanin (KLH). It has been found that the di- and tetrasaccharide (one repeating unit) conjugates contain already epitopes capable of inducing 6B-specific, fully protective antibodies in rabbits and mice, respectively (Jansen et al. 2001). In another investigation of a vaccine against *Shigella dysenteriae* type 1, a series of specific, oligosaccharides was prepared on the basis of the tetrasaccharide repeating unit of the O-specific polysaccharide. The oligosaccharides were coupled to human serum albumin and the conjugates were applied as vaccines (Pozsgay et al. 1999).

Development of anti parasite vaccines is more complicated than for vaccines directed against encapsulated bacteria. The principle of attacking the invaders on the basis of unique cell surface carbohydrates is in principle universally applies. The parasites are genetically and biologically complex organisms. Fundamental studies are needed to gain insight into the processes like the control of immune responses and the induction of appropriate immunological memory (Tarlton 2005). Malaria has affected approximately 300–500 million people in different countries. The malarial toxin produced by *Plasmodium falciparum* contains a carbohydrate moiety that could presumably be mimicked to create a vaccine. The vaccine is based on a class of glycolipids called glycosyl-phosphatidylinositols (GPIs). They are a key to malarial activity and vaccines based on these molecules could trigger anti- GPI antibodies to neutralize the toxin. Mice immunized with *P. falciparum* GPI showed a high degree of protection as compared with control group and this vaccine will be soon tested in monkeys.

AIDS is a major disease affecting human populations across different nations. HIV type 1 (HIV-1) virus is responsible for the current global pandemic of HIV and AIDS. Carbohydrates, which are having a strong defense against host immune attack, can serve as targets for vaccines. An antigen for AIDS, gp120, which is a glycoprotein of viral envelope of HIV, has been discovered. The viral surface presents a glycoprotein, gp120, which plays a prominent role in the penetration of the virus into cells of the immune system. gp120-associated molecules interact with CD4 proteins and chemokine receptors on T-lymphocytes, macrophages and dendritic cells to initiate internalization of HIV by the host cell. Cyanovirin and scytovirin from cyanobacteria binds the carbohydrate part of gp120 and people who make the human antibody called 2G12 can survive HIV infection for a long time. Cyanovirin-N (CV-N), an 11-kDa protein (composed of 101 amino acids consisting of two sequence repeats) originally purified from extracts of the cyanobacterium *N. ellipsosporum*. The elucidation of CV-N crystal structures revealed the existence of a domain-swapped dimer, with two primary carbohydrate-binding sites and two secondary carbohydrate-binding sites on opposite ends of the dimer (Barrientos et al. 2002). The carbohydrate-recognition sites have a binding geometry of high-mannose glycans, in particular $\alpha(1,2)$-linked mannose oligomers. A monomeric 13-kDa protein isolated from the unicellular freshwater bloom-forming cyanobacterium *Microcystis viridis* NIES-102 strain (*Microcystis viridis* lectin (MVL)) was also shown to be composed of two tandemly repeated homologous domains, with specificity for $\alpha(1,6)$- and possibly $\alpha(1,3)$-mannose oligomers. Its smallest target is a $Man_2GlcNAc_2$ tetrasaccharide core. Scytovirin (SVN), a 9.7-kDa peptide, with 95 amino acids, has most recently been isolated from the cyanobacterium *Scytonema varium* and was shown to have a pronounced affinity for $\alpha(1,2)$– $\alpha(1,6)$-mannose trisaccharide units. Both CV-N and SVN inhibit HIV infection in cell culture, at 50 % effective concentrations of 0.1 and 0.3 nM. The carbohydrate moiety of gp120 is synthesized and included into 2G12—type antibodies and conjugated to the carrier. This vaccine is presently undergoing clinical trial and quite promising.

Over the past few years, anticancer immunotherapy has emerged as a new and exciting area for controlling tumors. The targets in this case are host cells that have undergone mutations leading to uncontrolled cell division and the ability to invade other tissues. Another defining feature of cancer is altered glycosylation, including increased expression of certain glycans, called tumour-associated carbohydrate antigens (TACAs; Meezan et al. 1969). Immunization is carried out with tumor-associated antigens aiming at stimulating specific immune response against cancer cells. Carbohydrate antigens are potential targets for such immune interventions since they are exposed at the surface of tumor cells where they are hidden on normal cells. Several carbohydrate based vaccines are under development to treat cancer. These vaccines are based on carbohydrate happens conjugated to a protein carrier. Vaccines based on exposed core protein with major histocompatibility complex unrestricted epitopes, and carbohydrate structures are targets for the immunotherapy of cancers of epithelial origin. A vaccine formulated using sialyl-Tn (STn) has proven to be highly targeted—specific in human trials, and the induction of high anti-sTn antibody titres correlated with prolonged survival of breast cancer patients. Immune responses against carbohydrate antigens have been categorized as Thymus Independent (TI) in nature because they do not require cognate interactions between antigen specific B and T cells. These antigens are known to stimulate an antibody response in athymic mice. However, carbohydrate epitopes over-expressed on the surfaces of cancer cells may evoke a B cell response when introduced in an appropriate fashion to a host's immune system. The carbohydrate antigen Globo-H is a potential target for vaccine therapy. In this the complex carbohydrate molecule—Globo-H, the hexasaccharide has been synthesized, conjugated to keyhole limpet hemocyanin and administered with the immunologic adjuvant QS-21 as a vaccine for patients with prostate cancer who relapsed after primary therapies such as radiation or surgery. Another carbohydrate based vaccine called GMK contains ganglioside GM2. GMK vaccination induces AntiGM2 antibodies that target melanoma cells. The vaccine is currently in phase-3 trials for malignant melanoma. Mucinous proteins (MUC1) were adopted early on as a cancer vaccine target due to its expression in a wide variety of cancers (Cheever et al. 2009). However, most of these initial approaches relied on traditional vaccination means with unglycosylated epitopes (Gilewski et al. 2000). The Boons group applied their three-component approach to create a MUC1 vaccine consisting of a GalNAc-glycosylated MUC1 peptide, the T-helper epitope and the Pam3CysSK4 TLR agonist (Lakshminarayanan et al. 2012). This was the first example of a MUC vaccine that could elicit both humoral and cellular immunity leading to the high titer production of IgG against the epitope in mice.

Glycans are ubiquitous and the ability to understand and control their functions is going to be vital to pioneering future biological and therapeutic breakthroughs. The recent development of new tools and techniques to study and produce structurally defined carbohydrates has spurred renewed interest in the therapeutic applications of glycans. Carbohydrate antigens are important targets for the development of vaccines and pathogen detection strategies. Vaccines based on isolated glycan structures have helped to prevent deadly infectious diseases, and to approach the nearly complete eradication of certain pathogens. Modifying purified microbial carbohydrates through chemical synthesis or completely procuring glycan fragments through organic synthesis facilitated the antigen discovery process. An improved understanding of structural features that determine antigenicity is a prerequisite for efficient vaccine design. The identification of key epitopes that protect from a potential immune challenge or infection is a crucial milestone. Multidisciplinary approaches involving advanced oligosaccharide synthesis, glycan array screening, and biophysical and computational methods have been utilized by many groups to attain this goal. Sulfated exopolysaccharides synthesized by different cyanobacteria are heterogeneous and structurally different, which makes research very challenging. They can grow under controlled conditions, making chemical composition, structure, and rheological behavior of their SPS more stable along the several harvesting periods. These polymers have already proven their beneficial effects, but a lot is still to be done.

References

Agnihotri SA, Mallikarjuna NN, Aminabhavi TM (2004) Recent advances on chitosan-based micro- and nanoparticles in drug delivery. J Control Release 100:5–28

Alexandre KB, Gray ES, Lambson BE, Moore PL, Choge IA, Mlisana K, Karim SS, McMahon J, O'Keefe B, Chikwamba R, Morris L (2010) Mannose-rich glycosylation patterns on HIV-1 subtype C gp120 and sensitivity to the lectins, griffithsin, cyanovirin-*N* and scytovirin. Virology 402:187–196

Año G, Esquisabel A, Pastor M, Talavera A, Cedré B, Fernández S, Sifontes S, Aranguren Y, Falero G, García L et al (2011) A new oral vaccine candidate based on the microencapsulation by spray-drying of inactivated *Vibrio cholerae*. Vaccine 29:5758–5764

Ascencio F, Fransson LA, Wadström T (1993) Affinity of the gastric pathogen *Helicobacter pylori* for the N-sulfated glycosaminoglycan heparan sulfate. J Med Microbiol 38:240–244

Atkins EDT (1986) Biomolecular structures of naturally occurring carbohydrate polymers. Int J Biol Macromol 8:323–329

Avery OT, Goedel WF (1931) Chemo-immunological studies on conjugated carbohydrate-proteins: V the immunological specificity of an antigen prepared by combining the capsular polysaccharide of type III pneumococcus with foreign protein. J Exp Med 54:437–447

Axford J (2001) The impact of glycobiology on medicine. Trends Immunol 22:237–239

Barrientos LG, John ML, Istvan B, Toshiyuki M, Zhaozhong H, Barry RO, Michael RB, Alexander W, Angela MG (2002) The domain-swapped dimer of cyanovirin-N is in a metastable folded state: reconciliation of X-ray and NMR structures. Structure 10:673–686

Bellamy F, Horton D, Millet J, Picart F, Samreth S, Chana JB (1993) Glycosylated derivatives of benzophenone, benzhydrol, and benzhydril as potential venous antithrombotic agents. J Med Chem 36:898–903

Bergey E, Stinson M (1988) Heparin-inhibitable basement membrane-binding protein of *Streptococcus pyogenes*. Infect Immun 56:1715–1721

Bewley CA, Gustafson KR, Boyd MR, Covell DG, Bax A, Clore GM, Gronenborn AM (1998) Solution structure of cyanovirin-N, a potent HIV-inactivating protein. Nat Struct Biol 5:571–578

Bewley CA, Cai M, Ray S, Ghirlando R, Yamaguchi M, Muramoto K (2004) New carbohydrate specificity and HIV-1 fusion blocking activity of the cyanobacterial protein MVL: NMR, ITC and sedimentation equilibrium studies. J Mol Biol 339:901–914

Blinkova LP, Gorobets OB, Baturo AP (2001) Biological activity of *Spirulina* [In Russian]. Zhur Mikrobiol Epidemiol Immunobiol 5:114–118

Borchard G (2001) Chitosans for gene delivery. Adv Drug Deliv Rev 52:145–150

Buffa V, Stieh D, Mamhood N, Hu Q, Fletcher P, Shattock RJ (2009) Cyanovirin-N potently inhibits human immunodeficiency virus type 1 infection in cellular and cervical explant models. J Gen Virol 90:234–243

Buskas T, Li Y, Boons GJ (2004) The immunogenicity of the tumor-associated antigen Lewisy may be suppressed by a bifunctional cross-linker required for coupling to a carrier protein. Chemistry 10:3517–3524

Cameron RE (1962) Species of *Nostoc* vaucher occurring in the Sonoran Desert in Arizona. Trans Am Microsc Soc 81:379–384

Cassaro CMF, Dietrich CP (1977) Distribution of sulfated mucopolysaccharides in invertebrates. J Biol Chem 252:2254–2261

Cheever MA, Allison JP, Ferris AS, Finn OJ, Hastings BM, Hecht TT, Mellman I, Prindiville SA, Viner JL, Weiner LM, Matrisian LM (2009) The prioritization of cancer antigens: a National cancer institute pilot project for the acceleration of translational research. Clin Cancer Res 15:5323–5337

Chen B, You B, Huang J, Yu Y, Chen W (2010) Isolation and antioxidant property of the extracellular polysaccharide from *Rhodella reticulata*. World J Microbiol Biotechnol 26:833–840

Crouzier T, Boudou T, Picart C (2010) Polysaccharide-based polyelectrolyte multilayers. Curr Opin Colloid Interface Sci 15:417–426

Crowe LM (2002) Lessons from nature: the role of sugars in anhydrobiosis. Comp Biochem Phys A 131:505–513

Davis B (2002) Preparation of enzyme conjugated to carbohydrate moiety which binds to lectin-directed prodrug delivery system. PCT Int Appl Patent No. WO02/080980

De Philippis R, Vincenzini M (1998a) Exocellular polysaccharide from cyanobacteria and their possible applications. FEMS Microbiol Rev 22:151–175

DeMarco ML, Woods RJ (2008) Structural glycobiology: a game of snakes and ladders. Glycobiology 18:426–440

Dey B, Lerner DL, Lusso P, Boyd MR, Elder JH, Berger EA (2000) Multiple antiviral activities of cyanovirin-N: blocking of human immunodeficiency virus type 1 gp120 interaction with CD4 and coreceptor and inhibition of diverse enveloped viruses. J Virol 74:4562–4569

Duncan R (2003) The dawning era of polymer therapeutics. Nat Rev Drug Discov 2:347–360

Faroongsarng D, Sukonrat P (2008) Thermal behavior of water in the selected starch- and cellulose-based polymeric hydrogels. Int J Pharm 352:152–158

Filali Mouhim R, Cornet JF, Fontaine T, Fournet B, Dubertret G (1993) Production, isolation and characterization of the exopolysaccharide of the cyanobacterium *Spirulina platensis*. Biotechnol Lett 15:567–575

Fischetti L, Barry SM, Hope TJ, Shattock RJ (2009) HIV-1 infection of human penile explant tissue and protection by candidate microbicides. AIDS 23:319–328

Gilewski T, Adluri S, Ragupathi G, Zhang S, Yao T-J, Panageas K, Moynahan M, Houghton A, Norton L, Livingston PO (2000) Vaccina- tion of high-risk breast cancer patients with mucin-1 (MUC1) keyhole limpet hemocyanin conjugate plus QS-21. Clin Cancer Res 6:1693–1701

Gloaguen V, Morvan H, Hoffmann L (1995) Released capsular polysaccharides of *Oscillatoriaceae* (Cyanophyceae, Cyanobacteria). Algol Stud 78:53–69

Gloaguen V, Morvan H, Hoffmann L, Plancke Y, Wieruszeski JM, Lippens G, Strecker G (1999) Capsular polysaccharide produced by the thermophilic cyanobacterium *Mastigocladus laminosus*. Structural study of an undecasaccharide obtained by lithium degradation. Eur J Biochem 266:762–770

Gombotz WR, Wee S (1998) Protein release from alginate matrices. Adv Drug Deliv Rev 31:267–285

Guerrini M, Raman R, Venkataraman G, Torri G, Sasisekharan R, Casu B (2002) A novel computational approach to integrate NMR spectroscopy and capillary electrophoresis for structure assignment of heparin and heparan sulfate oligosaccharides. Glycobiology 12:713–719

Hayashi K, Hayashi T, Kojima IA (1996) Natural sulphated polysaccharide, calcium spirulan, isolated from *Spirulina platensis*: In vitro and ex vivo evaluation of anti-*herpes simplex* virus and anti-human immunodeficiency virus. AIDS Res Human Retrovir 12:1463–1471

He P, Davis SS, Illum L (1999) Chitosan microspheres prepared by spray drying. Int J Pharm 187:53–65

Hill DR, Keenan TW, Helm RF, Potts M, Crowe LM, Crowe JH (1997) Extracellular polysaccharide of *Nostoc* commune (cyanobacteria) inhibits fusion of membrane vesicles during desiccation. J Appl Phycol 9:237–248

Hoffman AS (2002) Hydrogels for biomedical applications. Adv Drug Deliv Rev 54:3–12

Huskens D, Ferir G, Vermeire K, Kehr JC, Balzarini J, Dittmann E, Schols D (2010) Microvirin, a novel alpha(1,2)-mannose-specific lectin isolated from *Microcystis aeruginosa*, has anti-HIV-1 activity comparable with that of cyanovirin-N but a much higher safety profile. J Biol Chem 285:24845–24854

Illum L, Jabbal-Gill I, Hinchcliffe M, Fisher AN, Davis SS (2001) Chitosan as a novel nasal delivery system for vaccines. Adv Drug Deliv Rev 51:81–96

Jansen WTM, Hogenboom S, Thijssen MJL, Kamerling JP, Vliegenthart JFG, Verhoef J, Snippe H, Verheul AFM (2001) Synthetic 6B di-,tri-, and tetrasaccharide-protein conjugates contain pneumococcal type 6 A and 6B common and 6B-specific epitopes that elicit protective antibodies in mice. Infect Immun 69:787–793

Jimenez-Barbero J, Asensio JL, Canada FJ, Poveda A (1999) Free and protein-bound carbohydrate structures. Curr Opin Struct Biol 9:549–555

Lakshminarayanan V, Thompson P, Wolfert MA, Buskas T, Bradley JM, Pathangey LB, Madsen CS, Cohen PA, Gendler SJ, Boons GJ (2012) Immune recognition of tumor-associated mucin MUC1 is achieved by a fully synthetic aberrantly glycosylated MUC1 tripartite vaccine. Proc Natl Acad Sci U S A 109:261–266

Li P, Harding SE, Liu Z (2002) Cyanobacterial exopolysaccharides: their nature and potential biotechnological applications. Biotechnol Genet Eng 18:375–404

Lipman CB (1941) The successful revival of *Nostoc* commune from a herbarium specimen eighty-seven years old. B Torrey Bot Club 68:664–666

Liu Z, Jiao Y, Wang Y, Zhou C, Zhang Z (2008) Polysaccharides-based nanoparticles as drug delivery systems. Adv Drug Deliv Rev 60:1650–1662

Mancuso Nichols CA, Guezennec J, Bowman JP (2005) Bacterial exopolysaccharides from extreme marine environments with special consideration of the southern ocean, sea ice, and deep-sea hydrothermal vents: a review. Mar Biotechnol 7:253–271

Marra M, Palmeri A, Ballio A, Segre A, Slodki ME (1990) Structural characterization of the exocellular polysaccharide from *Cyanospira capsulata*. Carbohyd Res 197:338–344

Meezan E, Wu HC, Black PH, Robbins PW (1969) Comparative studies on the carbohydrate-containing membrane components of normal and virus- transformed mouse fibroblasts. II. Separation of glycoproteins and glycopeptides by sephadex chromatography. Biochemistry 8:2518–2524

Micheletti E, Colica G, Viti C, Tamagnini P, De Philippis R (2008) Selectivity in the heavy metal removal by exopolysaccharide-producing cyanobacteria. J Appl Microbiol 105:88–94

Morais MG, Stillings C, Dersch R, Rudisile M, Pranke P, Costa JAV, Wendorff J (2010) Preparation of nanofibers containing the microalga *Spirulina (Arthrospira)*. Bioresour Technol 101:2872–2876

Moran AP, Khamri W, Walker MM, Thursz MR (2005) Role of surfactant protein D (SP-D) in innate immunity in the gastric mucosa: evidence of interaction with *Helicobacter pylori* lipopolysaccharide. J Endotoxin Res 11:357–362

Morelli L, Poletti L, Lay L (2011) Carbohydrates and Immunology: synthetic oligosaccharide antigens for vaccine formulation. Eur J Org Chem 2011:5723–5777

Mori T, O'Keefe BR, Sowder RC, Bringans S, Gardella R, Berg S, Cochran P, Turpin JA, Buckheit RW Jr, McMahon JB, Boyd MR (2005) Isolation and characterization of griffithsin, a novel HIV-inactivating protein, from the red alga *Griffithsia* sp. J Biol Chem 280:9345–9353

Mozzi F, Savoy de Giori G, Font de Valdez G (2003) UDP-galactose 4-epimerase: a key enzyme in exopolysaccharide formation by *Lactobacillus* casei CRL 87 in controlled pH batch cultures. J Appl Microbiol 94:175–183

Ni J, Song H, Wang Y, Stamatos NM, Wang LX (2006) Toward a carbohydrate-based HIV-1 vaccine: synthesis and immunological studies of oligomannose-containing glycoconjugates. Bioconjug Chem 17:493–500

Ofek I, Beachey EH, Sharon N (1978) Surface sugars of animal cells as determinants of recognition in bacterial adherence. Trends Biochem Sci 3:159–160

O'Keefe BR, Vojdani F, Buffa V, Shattock RJ, Montefiori DC, Bakke J, Mirsalis J, d'Andrea AL, Hume SD, Bratcher B, Saucedo CJ, McMahon JB, Pogue GP, Palmer KE (2009) Scaleable manufacture of HIV −1 entry inhibitor griffithsin and validation of its safety and efficacy as a topical microbicide component. Proc Natl Acad Sci U S A 106:6099–6104

Oh JK, Lee DI, Park JM (2009) Biopolymer-based microgels/nanogels for drug delivery applications. Prog Polym Sci 34:1261–1282

Okajima-Kaneko M, Ono M, Kabata K, Kaneko T (2007) Extraction of novel sulfated polysaccharides from *Aphanothece sacrum* (Sur.) Okada, and its spectroscopic characterization. Pure Appl Chem 79:2039–2046

Peppas NA, Bures P, Leobandung W, Ichikawa H (2000) Hydrogels in pharmaceutical formulations. Eur J Pharm Biopharm 50:27–46

Pereira S, Zille A, Micheletti E, Moradas-Ferreira P, De Philippis R, Tamagnini P (2009) Complexity of cyanobacterial exopolysaccharides: composition, structures, inducing factors, and putative genes involved in their biosynthesis and assembly. FEMS Microbiol Rev 33:917–941

Pozsgay V, Chu C, Pannell L, Wolfe J, Robbins JB, Schneerson R (1999) Protein conjugates of synthetic saccharides elicit higher levels of serum IgG lipopolysaccharide antibodies in mice than do those of the O-specific polysaccharide from *Shigella dysenteriae* type 1. Proc Natl Acad Sci U S A 96:5194–5197

Radonic A, Thulke S, Achenbach J, Kurth A, Vreemann A, König T, Walter C, Possinger K, Nitsche A (2010) Anionic polysaccharides from phototrophic microorganisms exhibit antiviral activities to *Vaccinia virus*. J Antivir Antiretrovir 2:51–55

Raman R, Raguram S, Venkataraman G, Paulson JC, Sasisekharan R (2005) Glycomics: an integrated systems approach to structure-function relationships of glycans. Nat Methods 2:817–824

Reeves PR, Hobbs M, Valvano MA, Skurnik M, Whitfield C, Coplin D, Kido N, Klena J, Maskell D, Raetz CRH, Rick PD (1996) Bacterial polysaccharide synthesis and gene nomenclature. Trends Microbiol 4:495–503

Richert L, Golubic S, Guedes RL, Ratiskol J, Payri C, Guezennec J (2005) Characterization of exopolysaccharides produced by cyanobacteria isolated from polynesian microbial mats. Curr Microbiol 51:379–384

Sato Y, Murakami M, Miyazawa K, Hori K (2000) Purification and characterization of a novel lectin from a freshwater cyanobacterium, *Oscillatoria agardhii*. Comp Biochem Physiol B Biochem Mol Biol 125:169–177

Schaeffer DJ, Krylov VS (2000) Anti-HIV activity of extracts and compounds from algae and cyanobacteria. Ecotoxicol Environ Saf 45:208–227

SjÖustrÖum JE, Larsson H (1996) Factors affecting growth and antibiotic susceptibility of *Helicobacter pylori*: effect of pH and urea on the survival of a wild-type strain and a urease-deficient mutant. J Med Microbiol 44:425–433

Skiba M, Morvan C, Duchene D, Puisieux F, Wouessidjewe D (1995) Evaluation of gastrointestinal behaviour in the rat of amphiphilic β-cyclodextrin nanocapsules, loaded with indomethacin. Int J Pharm 126:275–279

Sutherland IW (1994) Structure-function relationships in microbial exopolysaccharides. Biotechnol Adv 12:393–448

Sutherland IW (1998) Novel and established applications of microbial polysaccharides. Tibtech 16:41–46

Sutherland IW (2001) Microbial polysaccharides from gram-negative bacteria. Int Dairy J 11:663–674

Tarlton RL (2005) New approaches in vaccine development for parasitic infections. Cell Microbiol 7:1379–1386

Venkataraman G, Shriver Z, Raman R, Sasisekharan R (1999) Sequencing complex polysaccharides. Science 286:537–542

Volk RB, Venzke K, Blaschek W (2007) Structural investigation of a polysaccharide released by the cyanobacterium *Nostoc insulare*. J Appl Phycol 19:255–262

Wormald MR, Petrescu AJ, Pao YL, Glithero A, Elliott T, Dwek RA (2002) Conformational studies of oligosaccharides and glycopeptides: complementarity of NMR, X-ray crystallography, and molecular modelling. Chem Rev 102:371–386

Xiong S, Fan J, Kitazato K (2010) The antiviral protein cyanovirin-N: the current state of its production and applications. Appl Microbiol Biotechnol 86:805–812

Yoshimura H, Okamoto S, Tsumuraya Y, Ohmori M (2007) Group 3 sigma factor gene, sigJ, a key regulator of desiccation tolerance, regulates the synthesis of extracellular polysaccharide in cyanobacterium *Anabaena* sp. strain PCC 7120. DNA Res 14:13–24

Chapter 4
Anticancer Drug Development from Cyanobacteria

4.1 Cyanobacterial Metabolites with Potential Anticancer Properties

Cancer is medically known as a malignant neoplasm, which is a broad group of diseases involving unregulated cell growth. In cancer, cells divide and grow uncontrollably, forming malignant tumors, and invade nearby parts of the body. Cancer remains the second most common cause of death in the USA, accounting for nearly one of every four deaths. According to an estimate of the American Cancer Society, there will be an estimated 1,665,540 new cancer cases diagnosed and 585,720 cancer deaths in the USA in 2014. The most effective management of the cancer is surgical removal of the cancerous tissue followed by radiation therapy and/or chemotherapy. Cancer treatments do not have potent medicine as the currently available drugs are causing many side effects. There is an urgent need of new anticancer drugs because tumor cells are developing resistance against currently available drugs. In addition, the incidence of new types of cancer, such as the gliobastoma is increasing rapidly. Nature is the richest source of active principles against cancer cells. Natural compounds comprise either classical cytotoxic moieties targeting nonspecific macromolecules expressed by cancer cells and by normal proliferating cells (e.g., DNA, enzymes, microtubules), or new compounds targeting macromolecules specifically expressed on cancer cells (e.g., oncogenic signal transduction pathways). The medical treatment of cancer has made substantial improvements since the early years of modern antitumor drug research. A selected number of human malignancies (e.g., childhood lymphoblastic leukemia, lymphomas, and testicular cancer) can be cured with today's therapies and prolonged survival has been obtained in several others (Garattini and Vecchia 2001). The identification and development of natural compounds and their derivatives have greatly contributed to this progress and many of these compounds are now being used in clinical practice.

Cyanobacteria are considered a prominent source of structurally diverse and biologically active natural products (Nunnery et al. 2010). Cyanobacteria have a wide

© Springer International Publishing Switzerland 2015
S. Mandal, J. Rath, *Extremophilic Cyanobacteria For Novel Drug Development*,
SpringerBriefs in Pharmaceutical Science & Drug Development,
DOI 10.1007/978-3-319-12009-6_4

range of enzymes responsible for methylations, oxidations, tailorings, and other alterations (Jones et al. 2010), resulting in chemically diverse natural products such as linear peptides (Simmons et al. 2006), cyclic peptides (Sisay et al. 2009), linear lipopeptides (Nogle et al. 2001), depsipeptides (Han et al. 2005), cyclic depsipeptides (Soria-Mercado et al. 2009), fatty acid amides (Chang et al. 2011), swinholides (Andrianasolo et al. 2005), glicomacrolides (Teruya et al. 2009), or macrolactones (Salvador et al. 2010). More than 50 % of the marine cyanobacteria are potentially exploitable for extracting bioactive substances, which are effective in killing the cancer cells. Cryptophycins are potent anticancer agents produced by *Nostoc* sp. (Shih and Teicher 2001). It has an IC_{50} of 5 pg ml^{-1} against KB human nasopharyngeal cancer cells, and 3 pg ml^{-1} against LoVo human colorectal cancer cells, thus it was found to be 100–1000 times more potent than currently available anticancer drugs. It also exhibits anticancer activity against adriamycin-resistant M 17 breast cancer and DMS 273 lung cancer cell lines. The mechanism of anticancer activity of the cryptophycins has been associated with their destabilization of microtubules and with their induction of bcl-2 phosphorylation leading to apoptosis (Shih and Teicher 2001). The cryptophycins maintain activity against ovarian and breast carcinoma cells that overexpress the multidrug resistance efflux pump P-glycoprotein. An analogue cryptophycin 52 has demonstrated a broad range of antitumor activity against both murine and human tumors. Another potent antitumor agent dolastatin 10 was originally isolated from the Indian Ocean sea hare *Dolabella auricularia*. However, Luesch et al. (2001a) reported and isolated it from the marine cyanobacterium *Symploca* sp. from Palau. Dolastatin 10 is a linear pentapeptide consisting of a valine unit along with four unique residues including dolavaline, dolaisoleucine, dolaproline, and dolaphenine. Dolastatin 10 is the most active molecule in inhibiting cancer cell growth. The GI_{50} values (in µg ml^{-1}) of dolastatin 10 when tested in various human tumor cell lines including OVCAR-3, SF-295, A498, NCI-H460, KM20L2, and SK-MEL-5 were 9.5×10^{-7}, 7.6×10^{-6}, 2.6×10^{-5}, 3.4×10^{-6}, 4.7×10^{-6}, and 7.4×10^{-6}, respectively (Tan 2010). It binds to tubulin on the rhizoxin-binding site and affects microtubule assembly arresting the cell into G_2/M phase. The chemically related analogue symplostatin has been isolated from Guamanian and Hawaiian *Symploca hydnoides* is also a potent microtubule inhibitor and effective against a drug-insensitive mammary tumor and drug-insensitive colon tumor. Curacin A, isolated from *Lyngbya majuscula* is a potent inhibitor of cell growth and mitosis (Verdier-Pinard et al. 1998). Its structure is unique in that it contains the sequential positioning of a thiazoline and cyclopropyl ring, and it exerts its potential cell toxicity through interaction with the colchicine drug-binding site on microtubules. Currently, this compound is undergoing preclinical evaluation as potential future anticancer drugs. Apratoxin A is a cyclodepsipeptide isolated from a *Lyngbya* sp. collected from Guam displayed strong cytotoxicity against several cancer cell lines derived from colon, cervix, and bone. It is a mixed peptide-polyketide natural product of the Apratoxin family of cytotoxins, known for inducing G1-phase cell cycle arrests. Apratoxin A exhibited potent *in vitro* cytotoxicity in various human tumor cell lines with IC_{50} values ranging from 0.36 nM in LoVo cancer cells to 0.52 nM in KB cancer cells (Luesch et al. 2001b).

Tumor cells that survive initial chemotherapy in cancer patients often emerge with increased resistance to both the original therapeutic agent and seemingly unrelated drugs. This phenomenon is termed as multidrug resistance (MDR) and is often associated with increased expression of P-glycoproteins, which acts as an energy-dependent drug efflux pump. Tolyporphin was isolated from *Tolypothrix nodosa*, which exhibits reversal of MDR in a vinblastine resistance subline of a human ovarian adenocarcinoma cell line. It potentiates the cytotoxicity of Adriamycin or vinblastine in SK-VLB cells at doses as low as 1 µg/ml (Prinsep et al. 1992). It exhibits a potent photosensitizing activity against tumor cells and is 5000 times more effective than the photodynamic treatment (photofrin II; Morlière et al. 1998). Somocystinamide A (ScA), another extraordinary disulfide dimer of mixed PKS/ NRPS metabolites was isolated from the marine cyanobacterium *L. majuscula*. It is a pluripotent inhibitor of angiogenesis and tumor cell proliferation. It induces apoptosis in tumor and angiogenic endothelial cells. *In vitro* picomolar concentrations of ScA are sufficient to disrupt proliferation and tubule formation in endothelial cells (Wrasidlo et al. 2008). Coibamide A isolated from a *Leptolyngbya* strain collected in Coiba National Park (Coiba Island, Panama), demonstrated anticancer activity against lung cancer NCI-H460, breast cancer MDA-MB-231, melanoma LOX IMVI, leukemia HL-60, and astrocytoma SNB75 (Medina et al. 2008). Lagunamide A isolated from the filamentous marine cyanobacterium, *L. majuscula*, from Pulau Hantu, Singapore, exhibited a selective growth inhibitory activity against a panel of cancer cell lines, including P388, A549, PC3, HCT8, and SK-OV3 cells, with IC_{50} values ranging from 1.6 to 6.4 nM (Tripathi et al. 2012). Screening of metabolites derived from *L. symploca* led to the discovery and structure elucidation of the promising anticancer agent largazole. Largazole is a macrocyclic depsipeptide that exhibits a selectively potent anticancer activity through the inhibition of histone deacetylases (HDAC; Cole et al. 2011). The normal biological activity of HDAC involved regulation of transcription in eukaryotic cells. Therefore HDAC inhibitors will induce cell differentiation and cell death, thus blocking the proliferation of tumor cells and is a promising anticancer drug candidates (Wang et al. 2011).

Multiple myeloma (MM) is the second most common hematologic malignancy (Cruz et al. 2011). The disease is characterized by the accumulation of mature antibody-producing plasma cells in the bone marrow (Lamelin and Vassalli 1978). Despite the recent advances in the therapy of MM and the emergence of novel agents such as bortezomib, thalidomide, and lenalidomide (Muta et al. 2013; Senthilkumar and Ganesh 2012), the disease remains incurable. Scytonemin, isolated from cyanobacteria (Proteau et al. 1993), is able to inhibit the proliferation of three myeloma cells in a dose-dependent manner. Scytonemin is the first described small molecule inhibitor of human polo-like kinase. It inhibited the Plk1 activity in a concentration-dependent manner with an IC_{50} of 2 mM against the recombinant enzyme (Stevenson et al. 2002). Scytonemin induced the inhibition of cell growth and cell cycle arrest in multiple myeloma cells by specifically decreasing Plk1 activity and is promising novel agent for the treatment of multiple myeloma. In another study, Itoh et al. (2013) found reduced-scytonemin (R-scy) isolated from *Nostoc commune*, which has been shown to suppress the human T-lymphoid Jurkat cell growth.

The protein p53 is a well-characterized tumor suppressor that acts as a transcription factor to regulate cell cycle dynamics, apoptosis, and DNA repair. Hoiamide D isolated from Papua New Guinea cyanobacterium *Symploca* sp. displays promising inhibitory activity toward p53/MDM2 interaction (EC_{50} = 4.5 lM) and an attractive target for anticancer drug development (Malloy et al. 2012). Another class of marine cyclic depsipeptides having actin disruption activity is the lyngbyabellins/ hectochlorins isolated from *L. majuscula* (Marquez et al. 2002). These are unique bithiazole-containing compounds having the unusual dichlorinated β-hydroxyl acid residue, 7,7-dichloro-3-hydroxy-2-methyl-octanoic acid. The pharmacological target for hectochlorin was suggested to be actin microfilaments due to accumulation of CA46 cells in the G2/M phase of the cell cycle and induces actin polymerization in PtK2 cells with EC_{50} value at 20 μM. Biological evaluation of hectochlorin in the NCI 60 cancer cell lines showed significant cytotoxicity against a number of cancer cell lines, including colon melanoma, ovarian, and renal cells. In addition, the dose–response curve of hectochlorin also suggests that the molecule is antiproliferative instead of cytotoxic (Marquez et al. 2002). Many cyanobacterial compounds target specific macromolecules/enzymes that are implicated in cell proliferation. A high proportion of these compounds disrupt the normal functions of microtubules/actin microfilaments in eukaryotic cells, making these molecules invaluable as potential anticancer drugs.

4.2 Cytotoxicity of Cyanobacterial Natural Product

Cytotoxic compounds in general are substances inhibiting or stopping the growth of individual cells. In most cases, these compounds usually exhibit a wide range of effects on unicellular organisms and invertebrates (Biondi et al. 2004). Cyanobacterial metabolites exhibit a wide range of biological effects. Some of these metabolites possess potential cytotoxic activities to different mammalian cell lines, making them interesting for pharmaceutical research and application. In general, cytotoxins are very heterogeneous regarding their chemical structures and modes of action. The inductions of oxidative stress, actin depolymerization, or interference with the DNA replication process are the most commonly described effects for cyanobacterial cytotoxins. Most of these compounds affect a wide range of organisms. Tolytoxin, one of the most effective cytotoxin, causes profound injury to nearly all eukaryotic organisms, including fungi and different invertebrates (Patterson and Carmeli 1992). A similar wide-ranging effect can also be found in other cytotoxins such as calothrixins, cryptophycins, and pahayakolides (Berry et al. 2004; Biondi et al. 2004). Cylindrospermopsin (CYN) is produced by several freshwater cyanobacteria, such as *Cylindrospermopsis raciborskii*, *Umezakia natans*, *Aphanizomenon ovalisporum*, *Rhaphidiopsis curvata*, and *Anabaena bergii*. Cylindrospermopsin (CYN) is a tricyclic guanidine derivative containing a hydroxymethyluracil group and classified as hepatotoxins since exposure causes the greatest damage to the liver/hepatopancreas. Exposure to CYN results in the irreversible inhibition of protein synthesis, cellular necrosis, atrophy, DNA fragmentation, and tumor initiation or mutagenesis (Humpage et al. 2005). CYN is known to inhibit the division of animal cells and the protein synthesis in eukaryotes, although the molecular mechanism is yet to be understood (Metcalf et al. 2002). The alkaloid probably exerts its effect through DNA intercalation followed by strand cleavage. Cyclodepsipeptides, lagunamides A and B, isolated from *L. majuscula* show significant cytotoxicity having IC_{50} values of 6.4 and 20.5 nM, respectively, when tested against the P388 murine leukemia cell line (Tripathi et al. 2010).

Several genera of cyanobacteria such as *Microcystis, Anabaena, Nostoc, Oscillatoria, Cylindrospermum, Cylindrospermopsis, Aphanizomenon*, and *Nodularia* are known to produce toxins such as hepatotoxins, neurotoxins, cytotoxins, and dermatotoxins. Among these toxins, cyanobacterial cyclopeptides microcystins and nodularins are considered a danger to human health because of the possible toxic effects of high consumption (Soni et al. 2008). Although cyanotoxins can cause damage, they are also considered a rich source of natural cytotoxic compounds with potential anticancer action. From the pharmacological point of view, microcystins are stable hydrophilic cyclic heptapeptides with the potential to cause cellular damage following uptake via organic anion transporting polypeptides (OATPs; Uzair et al. 2012). Their biological effects involve intracellular inhibition of the catalytic subunit of protein phosphatase 1 (PP1) and PP2, glutathione depletion, and generation of reactive oxygen species (ROS; Gupta et al. 2003). Cancer cells live in a state of increased basal oxidative stress, which makes them vulnerable to further ROS insults induced by exogenous agents. Therefore, microcystin analogues can selectively kill cancer cells that express (AOTP) without causing significant toxicity to normal cells by exploiting the differences between normal and cancer cell redox (Sainis et al. 2010). Since OATPs are prominently expressed in cancers as compared with normal tissues, microcystins may be potential candidates for the development of anticancer drugs (Uzair et al. 2012). Cyanotoxins include a rich source of natural cytotoxic compounds with the potential to target cancers by expressing specific uptake transporters. Cyanobacteria of pharmacological and toxicological significance include the genera *Anabaena, Oscillatoria, Microcystis, Nodularia, Cylindrospermopsis*, and *Lyngbya*. These cyanobacteria produce diverse structural classes of metabolites such as acyclic peptides, linear decadepsipeptides, linear lipopeptides, linear alkynoic lipopeptides, cyclic depsipeptides, cyclic undecapeptides, cyclic hexapeptides, lipophylic cyclic peptides, paracyclophanes, sesterterpenes, glycolipids, polyphenolic ethers, and macrolactones, and most of them show cytotoxicity to various cancer cell line (Nagarajan et al. 2012). Cryptophycin was isolated in the laboratory from *Nostoc* sp. GSV224, and showed action on nasopharyngeal cancer cells and human colorectal cancer cells, being found to be 100–1000 times more potent than the anticancer drugs currently available, such as taxol or vinblastine (Trimurtulu et al. 1994). Cytotoxicity of the cyanobacteria *Oscillatoria* spp. were studied by Nair and Bhimba (2013), and the results indicated that *Oscillatoria boryana* showed potent cytotoxic activity against cell lines of human breast cancer. *Oscillatoria margaritifera* produce a cyclodepsipeptide veraguamides A from dichloromethane and methanol extract and showed potent cytotoxicity to the H-460

(LD$_{50}$ 141 nM) human lung cancer cell line (Mevers et al. 2011). Meickle et al. (2009) studied potential cytotoxic compounds from *Lyngbya* sp. and it exhibited effective cytotoxic activity against KB cells, HT-29 colorectal adenocarcinoma and IMR-32 neuroblastoma cells. Cytotoxicity-guided isolation from *L. confervoides* gives grassypeptolide, which showed activity against HeLa cervical carcinoma with IC$_{50}$ value of 1.0 mM (Kwan et al. 2008). Cocosamides A was also isolated from *L. majuscula* and reported cytotoxic activity against MCF7 and HT-29 cell line with IC$_{50}$ 29 and 30 mM, respectively (Gunasekera et al. 2011). Cyclic depsipeptide of obyanamide and ulongapeptin were identified from *L. confervoides* and *Lyngbya* sp. and showed cytotoxic effects on KB cell with IC$_{50}$ values of 0.58 mg ml^{-1} and 0.63 mM, respectively (Williams et al. 2002, 2003b). *Symploca* is another potential cyanobacteria with many cytotoxic compounds. Micromide and guamamide isolated from *Symploca* sp., exhibited good *in vitro* cytotoxicity against KB cells with IC$_{50}$ values of 260 and 1200 nM (Williams et al. 2004). Belamide A also from *Symploca* sp. showed potent cytotoxic activity against HCT-116 colon cancer cell line at IC$_{50}$ value of 0.74 mM (Simmons et al. 2006). Tasipeptins A and B isolated from the same species also exhibited potent cytotoxicity against KB cells with IC$_{50}$ values of 0.93 and 0.82 mM (Williams et al. 2003a).

Approximately 30% of cyanobacterial extracts have been reported to cause damage to mammal cells *in vitro* (Surakka et al. 2005). These effects can be caused due to the presence of specific toxins affecting the cell metabolism or by the presence of more compounds with a synergic effect. The cytotoxicity of cyanobacterial natural compounds in cancer cell lines is induced by different mechanisms. Only in case of few cyanobacterial cytotoxins the mode of action has been explained. The interaction of a cyanobacterial secondary metabolite with cytoskeletal structures (e.g., cryptophycin and tolytoxin), different eukaryotic enzymes (microcystin, nodularin, abaenopeptins, microviridins, and aeruginosins), as well as DNA (e.g., tubercidin), has been proved (Barchi et al. 1983; MacKintosch et al. 1990; Murakami et al. 1997; Patterson et al. 1993; Smith et al. 1994). The most extensively studied compound, lyngyabellin A, was shown to disrupt the microfilament network, and accordingly to disrupt cytokinesis in colon carcinoma cells, causing the formation of apoptotic bodies (Han et al. 2005; Yokokawa et al. 2001). Coibamide, a cyclic depsipeptide, isolated from a Panamanian *Leptolyngbya* sp. strain, causes cell cycle arrest in the G$_1$ phase in MDA-MB-435 breast cancer cells (Medina et al. 2008). Bouillomides A and B, another two depsipeptides isolated from *L. bouillonii*, were found to specifically and strongly inhibit serine proteases elastase and trypsin (Rubio et al. 2010). Other marine cyanobacteria compounds are capable of inducing oxidative stress and DNA fragmentation, microfilament disruption, Bcl-2 protein family modulation and even alterations in cell membrane dynamics (Chen et al. 2003; Luesch et al. 2000; Kalemkerian et al. 1999; LePage et al. 2005). The high estimated number of cytotoxic compounds agrees well with the diversity in cytotoxic structures that have been described by various authors (e.g., Barchi et al. 1983; Patterson and Carmeli 1992; Rodney et al. 1999; Trimurtulu et al. 1994). Based on the potential of its synthetic apparatus, cytotoxic metabolites are probably synthesized by the cyanobacterium with respect to their cytotoxic function, and is a novel candidate for many such cytotoxic agents of pharmaceutical importance.

4.3 Apoptosis and Cyanobacterial Natural Product

Apoptosis or programmed cell death (PCD) is an important physiological process for maintaining homeostasis that allow proper development and remodeling of normal tissues, generating immune responses, and destroying abnormal cells. During apoptosis living cells go through a predictable series of events in which enzymes break them down internally and then fall apart. Apoptosis is the most common type of cell death; characterized as chromatin condensation, nuclear fragmentation, cell shrinkage, and membrane blebbing. It's deregulation, i.e., either loss of proapoptotic signals or gain of antiapoptotic signals, can lead to a variety of pathological conditions such as cancer initiation, promotion, and progression or results in treatment failures. As apoptosis does not usually trigger inflammatory or immune response, it becomes a preferable way of cancer cell death during cancer treatments. Apoptosis can be induced in cancer cells artificially, by treating tumors with drugs or radiation. This breakthrough finding has led to the idea that a completely novel way of cancer therapy might be developed using drugs that directly switch on the cell death machinery in tumors. Several anticancer drugs work as apoptotic modulators, to eliminate silent and cleanly the unwanted cells (Zhang 2002; Fischer and Schulze-Osthoff 2005). Apoptosis can be induced by both intrinsic and extrinsic signals, by multiple agents, as the natural flavonoid quercetin, the representative reactive oxygen species H_2O_2 (Mao et al. 2006) or even the UV radiation (Lytvyn et al. 2010).

Cyanobacteria produce a wide range of compounds that revealed apoptotic properties. Martins and coworkers demonstrated that HL-60 cells exposed to aqueous extracts of *Synechocystis* sp. and *Synechococcus* sp. strains, presented cell shrinkage showing that cells were developing apoptosis, and membrane budding, that occurs when the cell is fragmented into apoptotic bodies (Martins et al. 2008). Apoptotic cells also develop nuclear alterations, visible as nuclear fragmentation and chromatin condensation (Elmore 2007). Biselyngbyaside, a macrolide glycoside produced by *Lyngbya* sp., was found to induce apoptosis in mature osteoclasts, revealed by nuclear condensation (Yonezawa et al. 2012). Marine benthic *Anabaena* sp. extracts were found to induce apoptosis in acute myeloid leukemia cell line, with cells showing several described typical morphological markers, such as chromatin condensation, nuclear fragmentation, surface budding, and release of apoptotic bodies (Oftedal et al. 2010). Calothrixins isolated from *Calothrix* sp. induce apoptosis in human cancer cells (Chen et al. 2003), presumably through their capacity to undergo oxidative cycling (Bernardo et al. 2007). The microcystins are highly active against the protein phosphatase 1 and 2A, which are essential for many signal transduction pathways of eukaryotic cells, such as PCD (apoptosis; Hooser 2000; Fischer et al. 2000). Oftedal et al. (2011) concluded that the apoptosis-inducing activity appears to be abundant in cyanobacteria, in particular of the genera *Microcystis* or

Anabaena, and is independent of geographical localization. Ultra-rapid apoptosis were observed in primary hepatocytes following microinjection with both Micro-cystin-LR (MC-LR) and nodularin, characterized by cytoplasmic shrinkage, chromatin condensation, membrane blebbing, and procaspase-3 cleavage (Fladmark et al. 1999). Žegura et al. (2006) have found that MC-LR induced DNA damage in HepG2 hepatoma cells related to decreased intracellular glutathione. In another study, they observed a significantly increased ratio of expression of bax to bcl-2 induced by MC-LR, which suggests that apoptosis in HepG2 cells proceeds *via* the mitochondrial pathway (Žegura et al. 2008). It has been reported that MC-LR covalently binds to cysteine residues of PP1 and PP2A (MacKintosh et al. 1995). As a result, inhibition of PP1 and PP2A leads to hyperphosphorylation of functional and cytoskeletal proteins and finally to cell apoptosis (Hooser 2000; Fischer et al. 2000). Liu et al. (2014) tested 40 cyanobacterial extracts and many of them show apoptosis-inducing activity in acute myeloid leukemia cells, and some of them were able to overcome typical features related to chemotherapy resistance. Exopolysaccharides from *Aphanothece halophytica* suppress cell growth and induce apoptosis in HeLa cell, a human cervical cancer cell line by targeting a master unfolded protein response (UPR) regulator Grp78. Grp78 further promotes the expression of CHOP and downregulates expression of survivin, which leads to activate mitochondria-mediated downstream molecules and p53-survivin pathway, resulting in caspase-3 activation and causing apoptosis (Ou et al. 2014). These findings provide important clues for further evaluating the potential potency of exopolysaccharides from *A. halophytica* for use in cancer therapy.

Apoptosis is mediated by the activation of different caspase cascades (Earnshaw et al. 1999). In mammals, there are two major signaling systems that result in the activation of caspases, i.e., the extrinsic death receptor pathway (Ozoren and El-Deiry 2003; Thorburn 2004) and the intrinsic mitochondrial pathway (Kroemer 2003; Gupta et al. 2009). Increasing evidence indicates that these two pathways are not isolated systems but, instead, that there are links between them. Cyanobacterial peptide ScA (Nogle and Gerwick 2002; Wrasidlo et al. 2008) and C-phycocyanin have been observed to display potent caspase-dependent proapoptotic activity in different cancer cells. ScA, a lipopeptide, was isolated from a *L. majuscula/Schizothrix* sp. assemblage of marine cyanobacteria (Nogle and Gerwick 2002). ScA stimulates apoptosis in a number of tumor cell lines. Biological evaluation revealed that ScA selectively induced apoptosis in cancer cell lines with caspase-8 expressions in the nanomolar range (Wrasidlo et al. 2008). Caspase-8 plays an important role in the progression of cell death, and the activation of this pathway by ScA offers an attractive target for tumor suppression. In addition, antiangiogenetic effects of ScA was observed based on the zebra fish model. Furthermore, low application of ScA at 100 pmol inhibited the growth of caspase-8-expressing tumor cells in an *in vivo* system using chick chorioallantoic membrane seeded with neuroblastoma tumors. C-phycocyanin (C-PC), a tetrapyrrole–protein complex isolated from the cyanobacteria *Agmenellum quadruplicatum*, *Mastigocladus laminosus* (Schirmer et al. 1986), and *Spirulina platensis*, induce the activation of proapoptotic gene and downregulation of antiapoptotic gene expression, then facilitate the transduction of apoptosis signals, resulting in the apoptosis of HeLa cells *in vitro*. Caspases 2, 3, 4, 6, 8, 9, and 10 were activated in C-PC-treated HeLa cells, suggesting that C-PC-induced apoptosis was caspase dependent (Zheng et al. 2013). C-phycocyanin consists of two subunits, designated as α and β, and each subunit bears phycocyanobilin (Mccarty 2007). C-phycocyanin induces apoptosis in leukemia K562 cells, which is accompanied by the release of cytochrome c, a change in the Bcl-2/Bax ratio, and the cleavage of poly-ADP-ribose polymerase (Subhashini et al. 2004). The β subunit caused apoptosis in squamous cell carcinoma 686LN-M4C1 cells by depolymerizing microtubules and microfilaments, activating caspase-3 and caspase-8, and reducing the nuclear level of glyceraldehyde-3-phosphate dehydrogenase (Wang et al. 2007). The α subunit (CpcA) induced apoptosis in COLO 205 cells through the mitochondrial pathway (Lu et al. 2011). They found caspase-9 activity was markedly increased after CpcA treatment and caspase-8 activity was unchanged, indicating that CpcA-induced apoptosis through the intrinsic pathway.

There are several types of PCD, including apoptosis (PCD I), autophagy (PCD II), and necrosis (PCD III; Galluzzi et al. 2012). In a recent study, reduced-scytonemin (R-scy), isolated from *N. commune*, has been shown to suppress the human T-lymphoid Jurkat cell growth (Itoh et al. 2013). Their result suggests R-scy induced Jurkat cell growth inhibition is attributable to the induction of type II PCD (PCD II; autophagic cell death or autophagy). The cells treated with R-scy produced large amounts of reactive oxygen species (ROS), leading to the induction of mitochondrial dysfunction and ROS formation plays a critical role in R-scy-induced PCDII. In another study, Stevenson et al. (2002) through [^{3}H] thymidine and cell counts demonstrated scytonemin's concentration-dependent reduction in Jurkat T cell proliferation, but with no chemical toxicity detected via trypan blue exclusion tests. However, flow-cytometric analysis revealed that scytonemin displayed no ability to arrest the cells at any one phase of the cycle, but was able to induce cells to undergo apoptosis independent of cell cycle phase. This induction of apoptosis is a significant finding in that, unlike necrosis or chemical toxicity, scytonemin interrupts essential biochemical processes, which triggers the cell to undergo PCD. Calcium ions (Ca^{2+}) function as messengers in the regulation of many pathways, including cell cycle and apoptosis (Trump and Berezesky 1995). Antillatoxin, kalkitoxin, alotamide A ($C_{32}H_{49}O_5N_3S$), hoiamide B ($C_{45}H_{73}N_5O_{10}S_3$), etc., of *Lyngbya* spp. are known to target calcium channels (LePage et al. 2005; Soria-Mercado et al. 2009) and induce apoptosis.

A potential problem in bioprospecting is the loss of biosynthetic capability for the substances of interest upon culturing of the producing organism. Oftedal et al. (2011) noted apoptogenic activity in cyanobacterial strains maintained in culture for several years. *Phormidium* CYA 76, isolated in 1956, exhibited apoptosis-inducing activity and *Microcystis* CYA 264, isolated in 1990, produced substantial amounts of microcystin. So, compared the apoptogenicity of old and recent cultures, it shows no indications that the recently collected strains contained more apoptogens than the formerly collected. This suggests that the cyanobacteria maintain their original biosynthetic capability after isolation and culturing, and the cyanobacterial metabolites are one of the ideal candidates for induction of apoptosis for cancer treatment.

4.4 Drug Delivery System with Cellular Targets

Cancer can be treated by surgery, chemotherapy, radiation therapy, hormonal ther-apy, and targeted therapy, including immunotherapy such as monoclonal antibody therapy. The choice of therapy depends upon the location and grade of the tumor and the stage of the disease, as well as the general state of the patient. Complete re-moval of the cancer cells without damage to the rest of the body is the goal of treat-ment. Sometimes this can be accomplished by surgery, but the propensity of cancers to invade adjacent tissue or to spread to distant sites by microscopic metastasis often limits its effectiveness; chemotherapy and radiotherapy can have a negative effect on normal cells. The major problem with chemotherapy for the treatment of cancer is the lack of the selective toxicity, which results in a narrow therapeutic index, and thereby compromises clinical prognosis. To reduce damage to normal tissues, sub-optimal doses of anticancer chemotherapeutics are often administered. Additional-ly, the high interstitial fluid pressure of solid tumors results in poor biodistribution and penetration of drugs. It has been shown that the amount of drug accumulated in normal viscera is ~ 10–20-fold higher than that in the same weight of tumor site and many anticancer drugs are not able to penetrate more than 40–50 μm from the vas-culature (Hambley and Hait 2009; Minchinton and Tannock 2006). These defects often lead to incomplete tumor response, multiple drug resistance, and ultimately, therapeutic failure (Szakacs et al. 2006). Nanotechnology holds significant promise for circumventing these challenges, by enabling large amounts of therapeutic drugs to be encapsulated into nanoparticles. This simultaneously increases the half-life and reduces toxic adverse effects of drugs, improving their pharmacokinetic profile and therapeutic efficacy (Davis et al. 2008; Shi et al. 2010). Nanoparticles can pro-vide a controlled and targeted way to deliver the encapsulated anticancer drugs and thus result in high efficacy and low side effects. Nanoparticles, with their small size and appropriate surface coating, may have the ability to overcome the drug resis-tance problem and thus greatly improve the efficacy of chemotherapy. Anticancer therapy using multifunctional nanomaterial, can target a cancer or tumor, deliver therapeutic drugs, and monitor the tumor tissues.

Lee et al. (2014) prepared a photoluminescent carbon nanotags (green carbon nanotags, G-tags) from harmful cyanobacteria, which is very efficient for drug de-livery and imaging in cancer cells. The G-tags possess high solubility, excellent photo stability under long-term aqueous conditions up to 240 h and low cytotoxicity (<2 mg/ml). The highly oxygen-functionalized surface of the G-tags have a num-ber of functional groups (–OH, C=O, and O=C–O–H) that enable the attachment of anticancer drugs without any further modification, resulting in water-soluble properties and cancer cell uptake of G-tags. The doxorubicin-conjugated G-tags (T-tags;>0.1 mg/ml) induced death in cancer cells (HepG2 and MCF-7) *in vitro* at a higher rate than that of only G-tags. In the *in vivo* mice experiment, it also showed enhanced anticancer efficacy by T-tags at 0.01 mg/ml, indicating that the loaded doxorubicin retains its pharmaceutical activity. Their results indicate that these multifunctional T-tags can deliver doxorubicin to the targeted cancer cells and sense the delivery of doxorubicin by activating the fluorescence of G-tags from cyanobacteria.

Liposomes are the other most popular nanoparticles, which have been used for drug delivery. Liposomes are lipid bilayer vesicles, composed of either a single bilayer or multiple layers. They exist in our body for certain physiological functions. Liposomes have wide applications in drug delivery and have no biocompatibility problem. They can be applied for both hydrophilic drugs (to be included in their aqueous interior) and hydrophobic drugs (to be entrapped in the lipid bilayer). They release the drug rapidly, usually in a few hours and provide desired adhesion to, and friendly interaction with the biological cells. The targeting ability of liposomes can be acquired by ligand–acceptor interaction and other mechanisms (active targeting) or improved by surface modification using PEG, chitosan, lectin, peptide, polysaccharides, etc. (Feng and Chien 2003). In drug delivery, not only the physiological response of the body is relevant, but the release mechanism is also of much importance. In most cases, collagen systems are used for local drug delivery. Materials of biological origin frequently claim to be relatively safe, readily available, relatively inexpensive, and capable of specific interactions with cells (Geiger et al. 2003). Tissue-specific liposomes and collagens with the inclusion of cyanobacterial anticancer metabolites can be applied with minimum side effects and may be efficient for specific cellular target.

References

Andrianasolo EH, Gross H, Goeger D, Musafija-Girt M, McPhail K, Leal RM, Mooberry SL, Gerwick WH (2005) Isolation of swinholide A and related glycosylated derivatives from two field collections of marine cyanobacteria. Org Lett 7:1375–1378

Barchi JJ, Bortin TR, Furusawa E, Patterson GLM, Moore RE (1983) Identification of a cytotoxin from *Tolypothrix byssoidea* as tubericin. Phytochemistry 22:2851–2852

Bernardo PH, Chai CL, Le Guen M, Smith GD, Waring P (2007) Structure-activity delineation of quinones related to the biologically active calothrixin B. Bioorg Med Chem Lett 17:82–85

Berry JP, Gantar M, Gawley RE, Wang M, Rein KS (2004) Pharmacology and toxicology of pahayokolide A, a bioactive metabolite from a freshwater species of *Lyngbya* isolated from the Florida Everglades. Comp Biochem Physiol C 139:231–238

Biondi N, Piccardi R, Margheri MC, Rodolfi L, Smith GD, Tredici MR (2004) Evaluation of *Nostoc* strain ATCC 53789 as a potential source of natural pesticides. Appl Environ Microbiol 70:3313–3320

Chang TT, More SV, Lu IH, Hsu JC, Chen TJ, Jen YC, Lu CK, Li WS (2011) Isomalyngamide A, A–1 and their analogs suppress cancer cell migration in vitro. Eur J Med Chem 46:3810–3819

Chen X, Smith GD, Waring P (2003) Human cancer cell (Jurkat) killing by the cyanobacterial metabolite calothrixin A. J Appl Phycol 15:269–277

Cole KE, Dowling DP, Boone MA, Phillips AJ, Christianson DW (2011) Structural basis of the antiproliferative activity of largazole, a depsipeptide inhibitor of the histone deacetylases. J Am Chem Soc 133:12474–12477

Cruz RD, Tricot G, Zangari M, Zhan F (2011) Progress in myeloma stem cells. Am J Blood Res 1:135–45

Davis ME, Chen Z, Shin DM (2008) Nanoparticle therapeutics: an emerging treatment modality for cancer. Nat Rev Drug Discov 7:771–782

Earnshaw WC, Martins LM, Kaufmann SH (1999) Mammalian caspases: structure, activation, substrates, and functions during apoptosis. Annu Rev Biochem 68:383–424

Elmore S (2007) Apoptosis: a review of programmed cell death. Toxicol Pathol 35:495–516

Feng SS, Chien S (2003) Chemotherapeutic engineering: application and further development of chemical engineering principles for chemotherapy of cancer and other diseases. Chem Eng Sci 58:4087–4114

Fischer U, Schulze-Osthoff K (2005) Apoptosis-based therapies and drug targets. Cell Death Differ 12:942–961

Fischer WJ, Hitzfeld BC, Tencalla F, Eriksson JE, Mikhailov A, Dietrich DR (2000) Microcystin-LR toxicodynamics, induced pathology, and immunohistochemical localization in livers of blue-green algae exposed rainbow trout (*Oncorhynchus mykiss*). Toxicol Sci 54:365–373

Fladmark KE, Brustugun OT, Hovland R, Boe R, Gjertsen BT, Zhivotovsky B, Doskeland SO (1999) Ultrarapid caspase-3 dependent apoptosis induction by serine/threonine phosphatase inhibitors. Cell Death Differ 6:1099–1108

Galluzzi L, Kepp O, Trojel-Hansen C, Kroemer G (2012) Mitochondrial control of cellular life, stress, and death. Circ Res 111:1198–1207

Garattini S, La Vecchia C (2001) Perspectives in cancer chemotherapy. Eur J Cancer 37:128–47

Geiger M, Li RH, Friess W (2003). Collagen sponges for bone regeneration with rhBMP-2. Adv Drug Deliv Rev 55:1613–1629

Gunasekera SP, Owle CS, Montaser R, Luesch H, Paul VJ (2011) Malyngamide 3 and cocosamides A and B from the marine cyanobacterium *Lyngbya majuscula* from Cocos Lagoon, Guam. J Nat Prod 74:871–6

Gupta N, Pant SC, Vijayaraghavan R, Rao PV (2003) Comparative toxicity evaluation of cyanobacterial cyclic peptide toxin microcystin variants (LR, RR, YR) in mice. Toxicology 188: 285–296

Gupta, S, Kass GE, Szegezdi E, Joseph B (2009) The mitochondrial death pathway: a promising therapeutic target in diseases. J Cell Mol Med 13:1004–1033

Hambley TW, Hait WN (2009) Is anticancer drug development heading in the right direction? Cancer Res 69:1259–1262

Han B, Goeger D, Maier CS, Gerwick WH (2005) The wewakpeptins, cyclic depsipeptides from a Papua New Guinea collection of the marine cyanobacterium *Lyngbya semiplena*. J Org Chem 70:3133–3139

Hooser SB (2000) Fulminant hepatocyte apoptosis in vivo following microcystin-LR administration to rats. Toxicol Pathol 28:726–733

Humpage AR, Fontaine F, Froscio S, Burcham P, Falconer IR (2005) Cylindrospermopsin genotoxicity and cytotoxicity: role of cytochrome P-450 and oxidative stress. J Toxicol Env Heal A 68:739–753

Itoh T, Tsuzuki R, Tanaka T, Ninomiya M, Yamaguchi Y, Takenaka H, Ando M, Tsukamasa Y, Koketsu M (2013) Reduced scytonemin isolated from *Nostoc commune* induces autophagic cell death in human T-lymphoid cell line Jurkat cells. Food Chem Toxicol 60:76–82

Jones AC, Monroe EA, Eisman EB, Gerwick L, Sherman DH, Gerwick WH (2010) The unique mechanistic transformations involved in the biosynthesis of modular natural products from marine cyanobacteria. Nat Prod Rep 27:1048–1065

Kalemkerian GP, Ou XL, Adil MR, Rosati R, Khoulani MM, Madan SK, Pettit GR (1999) Activity of dolastatin 10 against small-cell lung cancer in vitro and in vivo: induction of apoptosis and bcl-2 modification. Cancer Chemother Pharmacol 43:507–515

Kroemer G (2003) Mitochondrial control of apoptosis: an introduction. Biochem Biophys Res Commun 304:433–435

Kwan JC, Rocca JR, Abboud KA, Paul VJ, Luesch H (2008) Total structure determination of grassypeptolide: a new marine cyanobacterial cytotoxin. Org Lett 10:789–92

Lamelin JP, Vassalli P (1978) Heterogeneity of the B cell subpopulation operationally defined by (a) differentiation antigen(s) common to MOPC 104E and mature IgM plasma cells. Immunology 35:885–888

Lee HU, Park SY, Park ES, Son B, Lee SC, Lee JW, Lee YC, Kang KS, Kim MI, Park HG, Choi S, Huh YS, Lee SY, Lee KB, Oh YK, Lee J (2014) Photoluminescent carbon nanotags from harmful cyanobacteria for drug delivery and imaging in cancer cells. Sci Rep 4:4665

LePage KT, Goeger D, Yokokawa F, Asano T, Shioiri T, Gerwick WH, Murray TF (2005) The neurotoxic lipopeptide kalkitoxin interacts with voltage-sensitive sodium channels in cerebellar granule neurons. Toxicol Lett 158:133–139

Liu L, Herfindal L, Jokela J, Shishido TK, Wahlsten M, Døskeland SO, Sivonen K (2014) Cyanobacteria from terrestrial and marine sources contain apoptogens able to overcome chemoresistance in acute myeloid leukemia cells. Mar Drugs 12:2036–2053

Lu W, Yu P, Li J (2011) Induction of apoptosis in human colon carcinoma COLO 205 cells by the recombinant alpha subunit of C-phycocyanin. Biotechnol Lett 33:637–644

Luesch H, Yoshida WY, Moore RE, Paul VJ, Mooberry SL (2000) Isolation, structure determination, and biological activity of lyngbyabellin A from the marine cyanobacterium *Lyngbya majuscula*. J Nat Prod 63:611–615

Luesch H, Moore RE, Paul VJ, Mooberry SL, Corbett TH (2001a) Isolation of dolastatin 10 from the marine cyanobacterium *Symploca* species VP642 and total stereochemistry and biological evaluation of its analogue symplostatin 1. J Nat Prod 64:907–910

Luesch H, Yoshida WY, Moore RE, Paul VJ, Corbett TH (2001b) Total structure determination of apratoxin A, a potent novel cytotoxin from the marine cyanobacterium *Lyngbya majuscula*. J Am Chem Soc 123:5418–5423

Lytvyn DI, Yemets AI, Blume YB (2010) UV-B overexposure induces programmed cell death in a BY-2 tobacco cell line. Environ Exp Bot 68:51–57

MacKintosch C, Beattie KA, Klumpp S, Cohen P, Codd GA (1990) Cyanobacterial microcystin-LR is a potent and specific inhibitor of protein phosphatases 1 and 2A from both mammals and higher plants. FEBS Lett. 264:187–192

MacKintosh RW, Dalby KN, Campbell DG, Cohen PT, Cohen P, MacKintosh C (1995) The cyanobacterial toxin microcystin binds covalently to cysteine-273 on protein phosphatase 1. FEBS Lett 371:236–240

Malloy KL, Choi H, Fiorilla C, Valeriote FA, Matainaho T, Gerwick WH (2012) Hoiamide D, a marine cyanobacteria-derived inhibitor of p53/MDM2 interaction. Bioorg Med Chem Lett 22:683–688

Mao YB, Song G, Cai QF, Liu M, Luo HH, Shi MX, Ouyang G, Bao SD (2006) Hydrogen peroxide-induced apoptosis in human gastric carcinoma MGC803 cells. Cell Biol Int 30:332–337

Marquez BL, Watts KS, Yokochi A, Roberts MA, Verdier-Pinard P, Jimenez JI, Hamel E, Scheuer PJ, Gerwick WH (2002) Structure and absolute stereochemistry of hectochlorin, a potent stimulator of actin assembly. J Nat Prod 65:866–871

Martins RF, Ramos MF, Herfindal L, Sousa JA, Skaerven K, Vasconcelos VM (2008) Antimicrobial and cytotoxic assessment of marine cyanobacteria—*Synechocystis* and *Synechococcus*. Mar Drugs 6:1–11

McCarty MF (2007) Clinical potential of *Spirulina* as a source of phycocyanobilin. J Med Food 10:566–570

Medina RA, Goeger DE, Hills P, Mooberry SL, Huang N, Romero LI, Ortega-Barria E, Gerwick WH, McPhail KL (2008) Coibamide A, a potent antiproliferative cyclic depsipeptide from the Panamanian marine cyanobacterium *Leptolyngbya* sp. J Am Chem Soc 130:6324–6325

Meickle T, Matthew S, Ross C, Luesch H, Paul V (2009) Bioassay-guided isolation and identification of desacetyl-microcolin B from *Lyngbya* cf. polychroa. Planta Med 75:1427–30

Metcalf JS, Lindsay J, Beattie KA, Birmingham S, Saker ML, Törökné AK, Codd GA (2002) Toxicity of cylindrospermopsin to the brine shrimp *Artemia salina* comparisons with protein synthesis inhibitors and microcystins. Toxicon 40:1115–1120

Mevers E, Liu W, Engene N, Mohimani H, Byrum T, Pevzner PA, Dorrestein PC, Spadafora C, Gerwick WH (2011) Cytotoxic veraguamides, alkynyl bromide-containing cyclic depsipeptides from the marine cyanobacterium cf. *Oscillatoria margaritifera*. J Nat Prod 74:928–936

Minchinton AI, Tannock IF (2006) Drug penetration in solid tumours. Nat Rev Cancer 6:583–592

Morlière P, Mazière JC, Santus R, Smith CD, Prinsep MR, Stobbe CC, Fenning MC, Golberg JL, Chapman JD (1998) Tolyporphin: a natural product from cyanobacteria with potent photosensitizing activity against tumor cells in vitro and in vivo. Cancer Res 58:3571–3578

Murakami M, Sun Q, Ishida K, Matsuda H, Okino T, Yamuguchi K (1997) Microviridins, elastase inhibitors from the cyanobacterium *Nostoc minutum* (NIES-26). Phytochemistry 45:1997–1202

Muta T, Miyamoto T, Fujisaki T, Ohno Y, Kamimura T, Henzan T, Kato K, Takenaka K, Iwasaki H, Eto T, Takamatsu Y, Teshima T, Akashi K (2013) Effect of bortezomib-based induction therapy on the peripheral blood stem cell harvest in multiple myeloma. Rinsho Ketsueki 54:109–16

Nagarajan M, Maruthanayagam V, Sundararaman M (2012) A review of pharmacological and toxicological potentials of marine cyanobacterial metabolites. J Appl Toxicol 32:153–185

Nair S, Bhimba BV (2013) Bioactive potency of cyanobacteria *Oscillatoria* spp. Int J Pharm Pharm Sci 5:611–612

Nogle LM, Gerwick WH (2002) Somocystinamide A, a novel cytotoxic disulfide dimer from a Fijian marine cyanobacterial mixed assemblage. Org Lett 4:1095–1098

Nogle LM, Okino T, Gerwick WH (2001) Antillatoxin B, a neurotoxic lipopeptide from the marine cyanobacterium *Lyngbya majuscula*. J Nat Prod 64:983–985

Nunnery JK, Mevers E, Gerwick WH (2010) Biologically active secondary metabolites from marine cyanobacteria. Curr Opin Biotechnol 21:787–793

Oftedal L, Selheim F, Wahlsten M, Sivonen K, Doskeland SO, Herfindal L (2010) Marine benthic cyanobacteria contain apoptosis-inducing activity synergizing with daunorubicin to kill leukemia cells, but not cardiomyocytes. Mar Drugs 8:2659–2672

Oftedal L, Skjaerven KH, Coyne RT, Edvardsen B, Rohrlack T, Skulberg OM, Døskeland SO, Herfindal L (2011) The apoptosis-inducing activity towards leukemia and lymphoma cells in a cyanobacterial culture collection is not associated with mouse bioassay toxicity. J Ind Microbiol Biotechnol 38:489–501

Ou Y, Xu S, Zhu D, Yang X (2014) Molecular mechanisms of exopolysaccharide from *Aphanothece halaphytica* (EPSAH) induced apoptosis in HeLa cells. PloS ONE 9:e87223

Ozoren N, El-Deiry WS (2003) Cell surface death receptor signaling in normal and cancer cells. Semin Cancer Biol 13:135–147

Patterson GM, Carmeli S (1992) Biological effects of tolytoxin (6-hydroxy-7-O-methyl-scytophycin b) a potent bioactive metabolite from cyanobacteria. Ach Microbiol 157:406–410

Patterson GML, Smith CD, Kimura LH, Britton BA, Carmeli S (1993) Action of tolytoxin on cell morphology, cytoskeletal organization, and actin polymerization. Cell Motil Cytoskeleton 24:39–48

Prinsep MR, Caplan FR, Moore RE, Patterson GM, Smith CD (1992) Tolyporphin, a novel multidrug resistance-reversing agent from the blue-green alga *Tolypothrix nodosa*. J Am Chem Soc 114:385–387

Proteau PJ, Gerwick WH, Garcia-Pichel F, Castenholz R (1993) The structure of scytonemin, an ultraviolet sunscreen pigment from the sheaths of cyanobacteria. Experientia 49:825–9

Rodney WR, Rothschild JM, Willis AC, Chazal NM, Kirk J, Saliba KJ, Smith GD (1999) Calothrixin A and B, novel pentacyclic metabolites from *Calothrix* cyanobacteria with potent activity against malaria parasites and human cancer cells. Tetrahedron 55:13513–13520

Rubio BK, Parrish SM, Yoshida W, Schupp PJ, Schils T, Williams PG (2010) Depsipeptides from a Guamanian marine cyanobacterium, *Lyngbya bouillonii*, with selective inhibition of serine proteases. Tetrahedron Lett 51:6718–6721

Sainis I, Fokas D, Vareli K, Tzakos AG, Kounnis V, Briasoulis E (2010) Cyanobacterial cyclopeptides as lead compounds to novel targeted cancer drugs. Mar Drugs 8:629–657

Salvador LA, Paul VJ, Luesch H, Caylobolide B (2010) a macrolactone from symplostatin1-producing marine cyanobacteria *Phormidium* spp. from Florida. J Nat Prod 73:1606–1609

Schirmer T, Huber R, Schneider M, Bode W, Miller M, Hackert ML (1986) Crystal structure analysis and refinement at 2.5 Å of hexameric C-phycocyanin from the cyanobacterium *Agmenellum quadruplicatum*: the molecular model and its implications for light-harvesting. J Mol Biol 188:651–676

Senthilkumar CS, Ganesh N (2012) Lenalidomide-based combined therapy induced alterations in serum proteins of multiple myeloma patient: a follow-up case report and overview of the literature. Exp Oncol 34:373–376

Shi J, Votruba AR, Farokhzad OC, Langer R (2010) Nanotechnology in drug delivery and tissue engineering: from discovery to applications. Nano Letters 10:3223–3230

Shih C, Teicher BA (2001) Cryptophycins: a novel class of potent antimitotic antitumor depsipeptides. Curr Pharm Des 7:1259–1276

Simmons TL, McPhail KL, Ortega-Barrı́a E, Mooberry SL, Gerwick WH (2006) Belamide A: a new antimitotic tetrapeptide from a Panamanian marine cyanobacterium. Tetrahedron Lett 47:3387–3390

Sisay MT, Hautmann S, Mehner C, Konig GM, Bajorath J, Gutschow M (2009) Inhibition of human leukocyte elastase by brunsvicamides A–C: cyanobacterial cyclic peptides. Chem Med Chem 4:1425–1429

Smith CD, Zhang XQ, Moorbery SL, Patterson GML, Moore RE (1994) Cryptophycin—a new antimicrotubule agent active against drug-resistant cells. Cancer Res 54:3779–3784

Soni B, Trivedi U, Madamwar D (2008) A novel method of single step hydrophobic interaction chromatography for the purification of phycocyanin from *Phormidium fragile* and its characterization for antioxidant property. Bioresour Technol 99:188–194

Soria-Mercado IE, Pereira A, Cao Z, Murray TF, Gerwick WH (2009) Alotamide A, a novel neuropharmacological agent from the marine cyanobacterium *Lyngbya bouillonii*. Org Lett 11:4704–4707

Stevenson CS, Capper EA, Roshak AK, Marquez B, Grace K, Gerwick WH, Jacobs RS, Marshall LA (2002) Scytonemin—a marine natural product inhibitor of kinases key in hyper proliferative inflammatory diseases. Inflamm Res 51:112–114

Subhashini J, Mahipal SVK, Reddy MC, Mallikarjuna RM, Rachamallu A, Reddanna P (2004) Molecular mechanisms in C-Phycocyanin induced apoptosis in human chronic myeloid leukemia cell line -K 562. Biochem Pharmacol 68:453–462

Surakka A, Sihvonen LM, Lehtimaki JM, Wahlsten M, Vuorela P, Sivonen K (2005) Benthic cyanobacteria from Baltic Sea contain cytotoxic *Anabaena*, *Nodularia* and *Nostoc* strains and an apoptosis-inducing *Phormidium* strain. Environ Toxicol 20:285–292

Szakacs G, Paterson JK, Ludwig JA, Booth-Genthe C, Gottesman MM (2006) Targeting multidrug resistance in cancer. Nat Rev Drug Discov 5:219–234

Teruya T, Sasaki H, Kitamura K, Nakayama T, Suenaga K (2009) Biselyngbyaside, a macrolide glycoside from the marine cyanobacterium *Lyngbya* sp. Org Lett 11:2421–2424

Tan LT (2010) Filamentous tropical marine cyanobacteria: a rich source of natural products for anticancer drug discovery. J Appl Phycol 22:659-676

Thorburn A (2004) Death receptor-induced cell killing. Cell Signal 16:139–144

Trimurtulu G, Ohtani I, Patterson GM, Moore RE, Corbett TH, Valeriote FA, Demchik L (1994) Total structures of cryptophycins, potent antitumor depsipeptides from the blue-green alga *Nostoc* sp. strain GSV 224. J Am Chem Soc 116:4729–4737

Tripathi A, Puddick J, Prinsep MR, Rottmann M, Tan LT (2010) Lagunamides A and B: cytotoxic and antimalarial cyclodepsipeptides from the marine cyanobacterium *Lyngbya majuscula*. J Nat Prod 73:1810–1814

Tripathi A, Fang W, Leong DT, Tan LT (2012) Biochemical studies of the lagunamides, potent cytotoxic cyclic depsipeptides from the marine cyanobacterium *Lyngbya majuscula*. Mar Drugs 10:1126–1137

Trump BF, Berezesky IK (1995) Calcium-mediated cell injury and cell death. FASEB J 9:219–228

Uzair B, Tabassum S, Rasheed M, Rehman SF (2012) Exploring marine cyanobacteria for lead compounds of pharmaceutical importance. Sci World J 1–10.

Verdier-Pinard P, Lai JY, Yoo HD, Yu J, Marquez B, Nagle DG, Nambu M, White JD, Falck JR, Gerwick WH, Day BW, Hamel E (1998) Structure-activity analysis of the interaction of curacin A, the potent colchicine site antimitotic agent, with tubulin and effects of analogs on the growth of MCF-7 breast cancer cells. Mol Pharmaco 53:62–76

Wang L, Pan B, Sheng J, Xu J, Hu Q (2007). Antioxidant activity of *Spirulina platensis* extracts by
    supercritical carbon dioxide extraction. Food Chem 105:36–41
Wang B, Huang PH, Chen CS, Forsyth CJ (2011) Total syntheses of the histone deacetylase inhibi-
    tors largazole and 2-epi-largazole: application of n-heterocyclic carbene mediated acylations in
    complex molecule synthesis. J Org Chem 76:1140–1150
Williams PG, Yoshida WY, Moore RE, Paul VJ (2002) Isolation and structure determination of
    obyanamide: a novel cytotoxic cyclic depsipeptide from the marine cyanobacterium *Lyngbya
    confervoides*. J Nat Prod 65:29–31
Williams PG, Yoshida WY, Moore RE, Paul VJ (2003a) Tasipeptins A and B: new cytotoxic depsi-
    peptides from the marine cyanobacterium *Symploca* sp. J Nat Prod 66:620–624
Williams PG, Yoshida WY, Quon MK, Moore RE, Paul VJ (2003b) Ulongapeptin: a cytotoxic cy-
    clic depsipeptide from a Palauan marine cyanobacterium *Lyngbya* sp. J Nat Prod 66:651–654
Williams PG, Yoshida WY, Moore RE, Paul VJ (2004) Micromide and guamamide: cytotoxic
    alkaloids from a species of the marine cyanobacterium *Symploca*. J Nat Prod 67:49–53
Wrasidlo W, Mielgo A, Torres VA, Barbero S, Stoletov K, Suyama TL, Klemke RL, Gerwick WH,
    Carson DA, Stupack DG (2008) The marine lipopeptide somocystinamide A triggers apoptosis
    via caspase 8. PNAS 105:2313–2318
Yokokawa F, Sameshima H, Shioiri T (2001) Total synthesis of lyngbyabellin a, a potent cytotoxic
    metabolite from the marine cyanobacterium *Lyngbya majuscula*. Tetrahedron Lett 42:4171–
    4174
Yonezawa T, Mase N, Sasaki H, Teruya T, Hasegawa S, Cha BY, Yagasaki K, Suenaga K, Nagai K,
    Woo JT (2012) Biselyngbyaside, isolated from marine cyanobacteria, inhibits osteoclastogen-
    esis and induces apoptosis in mature osteoclasts. J Cell Biochem 113:440–448
Žegura B, Lah TT, Filipič M (2006) Alteration of intracellular GSH levels and its role in microcys-
    tin-LR-induced DNA damage in human hepatoma HepG2 cells. Mutat Res 611:25–33
Žegura B, Zajc I, Lah TT, Filipič M (2008) Patterns of microcystin-LR induced alteration of the
    expression of genes involved in response to DNA damage and apoptosis. Toxicon 51:615–623
Zhang JY (2002) Apoptosis-based anticancer drugs. Nat Rev Drug Discov 1:101–102
Zheng L, Lin X, Wu N, Liu M, Zheng Y, Sheng J, Xiaofeng JI, Sun M (2013) Targeting cellular
    apoptotic pathway with peptides from marine organisms. Biochim Biophys Acta 1836:42–48

Chapter 5
Issues and Challenges of Drug Development from Cyanobacteria

5.1 Clinical Trial and Commercial Drug Development

Drug development is a complex and costly process that includes the collection, analysis, and reporting of data from human subjects under strict clinical trials. Pharmaceutical companies routinely submit clinical trial results, as well as data, to regulatory agencies for licensing and other promotional activities pertaining to a new drug. The drug-development process normally proceeds through various phases of clinical trials (phase 0 or preclinical, phases I, II, and III). The Food and Drug Administration (FDA) must approve each phase before the study can continue. Drugs are first tested on laboratory animals (preclinical phase) and then on healthy humans in further phases. In phases I, II, and III the number of subjects range from 20 to 80, a few dozen to 300, and several hundred to 3000 people, respectively. The National Regulatory Authority (NRA) approves a drug for use by the general population if it successfully passes through all the phases of clinical trial (Dixit and Suseela 2013). Marine cyanobacteria are a rich source of diverse bioactive natural products and to date, 569 natural products of marine cyanobacterial origin have been reported in MarinLit (Munro and Blunt 2009). However, only few natural products from them are in clinical or preclinical trials for the treatment of cancer, inflammation, and other diseases. Cyanobacterial metabolites and their analogs that are in different stages of clinical trial are curacin A (preclinical), desmethoxymajusculamide C (DMMC; preclinical), cryptophycins (phase I), cemadotin (phase I), dolastatin 10 (phase II), tasidotin (phase II), dehydrodidemnin B (phase II), synthadotin (phase II), and soblidotin (phase III).

Curacins are unique thiazoline-containing lipopeptides that inhibit microtubule assembly and they are potent competitive inhibitors of the binding of colchicine to tubulin (Blokhin et al. 1995). The Gerwick group isolated curacin A (Gerwick et al. 1994); curacins B and C (Yoo and Gerwick 1995); and curacin D (Márquez et al. 1998) from *Lyngbya majuscule*. Clinical development of curacin has been hindered

© Springer International Publishing Switzerland 2015
S. Mandal, J. Rath, *Extremophilic Cyanobacteria For Novel Drug Development*,
SpringerBriefs in Pharmaceutical Science & Drug Development,
DOI 10.1007/978-3-319-12009-6_5

due to its low water solubility, as it is unable to act during *in vivo* animal trials. Hence, curacin was withdrawn from the preclinical phase, but it served as a lead compound for the development of synthetic analogs, which are more water soluble (Wipf et al. 2004) and are now in the preclinical phase. Cytotoxicity-guided fractionation of the organic extract from a Fijian *L. majuscula* led to the discovery of DMMC as the active metabolite. DMMC demonstrated potent and selective antisolid tumor activity with an $IC_{50}=20$ nM against the HCT-116 human colon carcinoma cell line via disruption of cellular microfilament networks (Simmons et al. 2009). A linear form of DMMC was generated by base hydrolysis and the amino acid sequence was confirmed by mass spectrometry. Linearized DMMC was also evaluated in the biological assays and found to maintain potent actin depolymerization characteristics while displaying solid tumor selectivity equivalent to DMMC in the disk diffusion assay and now it is in preclinical trial. Cryptophycin 1 was isolated from *Nostoc* sp. in Moore's lab as an anticancer agent (Patterson et al. 1991). It has an IC_{50} of 5 pg ml^{-1} against KB human nasopharyngeal cancer cells, and 3 pg ml^{-1} against LoVo human colorectal cancer cells, and thus it was found to be 100–1000 times more potent than currently available anticancer drugs, for example, taxol or vinblastine. It also exhibits anticancer activity against adriamycin-resistant M 17 breast cancer and DMS 273 lung cancer cell lines. It is a highly potent suppressor of microtubule dynamics and blocks the cells in the G2/M phase. There are several analogs of cryptophycin either naturally isolated or chemically synthesized. The synthetic analog cryptophycin 52, which progressed to phase II clinical trials for the treatment of patients with platinum-resistant ovarian cancer, is based on cryptophycin 1, which was isolated from terrestrial cyanobacteria (D'agostino et al. 2006). Currently, two other analogs, cryptophycins 249 and 309, with improved stability and water solubility are being considered as second-generation clinical candidates (Liang et al. 2005). The enhanced efficacy of these compounds compared with that of Cr-52 (LY 355073) has empowered new efforts to move these compounds into clinical trials (Yousong et al. 2008).

The National Cancer Institute of US conducted phase I clinical trials of dolastatin 10, which progressed to phase II. However, due to its toxic effects, phase II clinical trials of dolastatin 10 failed to show significant anticancer activity. This finding resulted in the development of dolastatin 10 synthetic analogs such as soblidotin, usually with improved pharmacological and pharmacokinetic properties. The *in vitro* studies of soblidotin (or TZT −1027, auristatin PE) a synthetic analog of dolastatin 10, showed promising results against human colon adenocarcinomas and cleared phases I and II successfully and is now undergoing phase III clinical trials under the supervision of Aska Pharmaceuticals, Tokyo, Japan (Bhatnagar and Kim 2010). The antitumor activity of soblidotin was found to be superior to existing anticancer drugs, such as paclitaxel and vincristine (Watanabe et al. 2006) and is quite promising as a drug if it clears all the phases of clinical trial. The third-generation dolastatin 15 analog is tasidotin. Tasidotin is an antitumor agent and has cleared phase I trials (Mita et al. 2006) and is now undergoing phase II trials under the supervision of Genzyme Corporation, Cambridge, MA (Bhatnagar and Kim 2010). Cematodin (LU-103793), a water-soluble analog of dolastatin 15, which has a terminal benzylamine moiety in place of dolapyrolidone, retains high cytotoxicity *in vitro*. It was found to be effective in a phase I trial for treatment of breast cancer and other cancers by BASF Pharma (Varanasi, India), but a phase II trial was discontinued following unexpected results. Synthadotin, another analog derived from dolastatin 15, showed promising results in phase II clinical trials of inoperable, locally advanced, or metastatic melanoma (Ebbinghaus et al. 2004). All of these metabolites offer great opportunity and a platform for the development of novel drugs.

Hayashi et al. 1994 investigated the effect of *Spirulina* in mice and reported increased phagocytic activity and increased antigen production in the test animals under study. Qureshi and Ali 1996 reported increased phagocytic activity, increased antigen production, and increased natural killer cell-mediated antitumor activity in chicken for the same cyanobacterium (*Spirulina*). In a preliminary small clinical study, an increase of 13.6-fold in interferon and 3-fold in interleukin (IL)-1β and −4 was observed in human blood cells incubated with *Spirulina* extracts. Khan et al. 2005 reported that different products prepared from *Spirulina* influence immune systems in various ways such as increasing the phagocytic activity of macrophages, stimulating antibody and cytokine production, increasing accumulation of natural killer cells into tissues and activating T and B cells. In a clinical study in Korea, (Park et al. 2008) a significant rise in plasma IL-2 concentration and a significant reduction in IL-6 concentration in humans was observed after the consumption of *Spirulina* at home, 8 g day^{-1}, for 16 consecutive weeks. *Spirulina* was found to be safe and the FDA approved it as a food colorant in the USA.

Despite many potent biological activities, few cyanobacterial compounds have entered clinical trials. The major problem lies with the supply of material for clinical studies. However, with the development of improved culture techniques their adequate supply should be feasible soon. Further, some of the cyanobacterial metabolites whose mechanisms of action (MOA) were discovered as a result of the screening system (NCI 60 Panel) are now either just entering or about to enter clinical trials and hope to pass clinical trials to become a drug candidate. To date, there are seven therapeutic agents (four anticancer, one antiviral, one pain control, and one for hypertriglyceridemia) that derive from the marine environment (Mayer et al. 2010). Of these, two are actual chemical structures as isolated, and five others are synthetic agents that capture their chemical idea from a marine product. In many cases mollusks, sponges, and tunicates are the richest collected sources of these most valuable metabolites. However, the collected source has often been shown or is strongly suspected of harboring or feeding upon cyanobacteria that are the actual producers of the bioactive agent. In many cases cyanobacteria are the real metabolic jewels of the world's oceans, accounting for fully 80% of these clinical trials and approved pharmaceutical agents. The pharmaceutical potential of cyanobacteria deserves more scientific attention and interdisciplinary research. Further, to find novel compounds, cyanobacterial strains from still unexplored and extreme habitats should also be studied.

5.2   Challenges of Drug Development in Cyanobacteria

Natural products have played an enormous role in the development of modern medicines. In the areas of cancer and infectious diseases, 60 and 75 % of new drugs, respectively, originated from natural sources between 1981 and 2002 (Newman et al. 2003). Between 2001 and 2005, 23 new drugs derived from natural products were introduced for the treatment of disorders such as bacterial and fungal infections, cancer, diabetes, dyslipidemia, atopic dermatitis, Alzheimer's disease, and genetic diseases such as tyrosinaemia and Gaucher disease. Cyanobacteria, produces a surprisingly diverse array of metabolites and immerges as an ideal candidate for new drug discovery. Currently, some 569 natural products of marine cyanobacterial origin have been reported in MarinLit (Munro and Blunt 2009). However, there are many challenges in the development of drugs from them. The core technologies used to discover natural products for drug discovery have not evolved substantially. New technologies that could improve the natural product drug discovery effort have not advanced to the degree at which the discovery rate of natural product-derived drugs would meet the demands of the industry. However, there have been many advances recently to speed up the process of drug development from them.

Many cyanobacteria synthesizes the novel metabolites in small quantities which, often present as mixtures in extracts and require labor-intensive and time-consuming purification procedures. Obtaining adequate quantities for preclinical development requires large-scale reacquisition or fermentation that would have a substantial impact on the development timeline. Rediscovery of known compounds is a major problem with cyanobacteria when screening natural product libraries. Dereplication is the process by which the chemical and biological characteristics of the unknown compounds are compared with the chemical and biological characteristics of known compounds from the databases to eliminate those that have been identified previously. Some natural products from cyanobacteria were discovered more frequently than others during screening programs. As the number of described metabolites increased, so did the probability of rediscovering known. The time-consuming processes of dereplication and purification are not compatible with the present regime of "blitz" screening campaigns in which assay support is only available for a limited duration (3 months). Further, in many cases the cyanobacterial metabolites are often structurally complex. Modification of complex natural products using organic chemistry is frequently challenging. Medicinal and combinatorial chemists prefer not to work with these natural products because of the large size and complexity of the compounds, which have too many functional groups to protect. It is difficult to prepare many natural product analogs as synthetic chemicals in the same time. Due to all these challenges the progress in drug discovery from them is very slow.

So far only a tiny fraction of the microbial world has been explored and the exponential growth in the microbial genomic database, the outlook for discovering new biologically active natural products from cyanobacteria is quite promising. Most of the cyanobacterial molecules discovered during the past decades were based on research conducted mainly at the laboratories of Professors Richard

Moore (University of Hawaii) and William Gerwick (University of California, San Diego). A majority of these biomolecules were isolated from the filamentous Order Nostocales, especially members belonging to the genera *Lyngbya*, *Oscillatoria*, and *Symploca*. The locations of the collection sites were mainly from the tropics, including Papua New Guinea and the Pacific Islands, particularly Guam and Palau. An advantage of natural products research on marine cyanobacteria is the high discovery rate ($>95\%$) of novel compounds as compared to other traditional microbial sources. This is due largely to the unexplored nature of this group of microalgae (Uzair et al. 2012). Therefore, one of the key areas to further tap these microalgae for new drug discovery is the collection of cyanobacterial strains from unexplored localities, and the amenability of field-collected strains to laboratory culture is an important factor in the drug discovery process from them. Once they have been brought into pure culture, straightforward procedures are available to cultivate them in larger volumes, to chemically analyze the natural products and identify the compounds, as well as to optimize the production by strain selection and elaboration of the optimal physico-chemical conditions for production. This includes design and development of the fermentation process and selection of strains from a larger panel of similar strains that produce the desired compound as well as strain improvement by random or directed genetic manipulation. The culturability of cyanobacteria would ensure a constant supply of potent secondary metabolites for further biological evaluation and chemical and biosynthetic studies. However, as indicated by Ramaswamy et al. (2006), the culturing of marine cyanobacteria faces several significant challenges, including physical parameters, sampling techniques, and issues associated with slow growing filamentous strains. However, there are several reports of development of photobioreactors and strategies specifically for cyanobacterial mass cultivation and the efficient production of valuable products (Suh and Lee 2003, Kim and Lee 2005, Kim et al. 2006). High biomass yields can be achieved in a fermenter-based system with agitation, control of effective light supply, and flow of carbon dioxide. From a commercial perspective, cyanobacterial culture and harvesting of products from them have several advantages. Cyanobacteria are mostly photoautotropic, requiring only sunlight, water, and basic nutrients for growth. Further unrestricted availability of exogenous carbon makes large-scale culture of cyanobacteria in open ponds commercially cheap (Apt and Behrens 1999). One notable advantage of some benthic cyanobacterial species is that during their growth cycle they become positively buoyant, detach from the growth surface and form floating mats. This "self-harvesting" capability could be advantageous in commercial photobioreactor as it would greatly reduce dewatering costs.

In order to perform a competitive and successful search for new drug candidates a multidisciplinary research team is required and the complete set of methods should be available. First of all, it is essential to conserve cyanobacterial cultures and build up culture collections to keep the strains for further investigation and production. The cultivation offers a wide range of methods for stimulation of secondary metabolite production. The crucial demand is for establishing and maintaining substance libraries with natural products of high purity to meet the demand of high throughput screening procedures. Third, the availability of attractive panels with a broad range of bioassays is key to success in finding new drug candidates. The cultivation of secondary metabolite-producing cyanobacteria under standard laboratory conditions often reveals only small metabolic profiles. It is clear from the genome information that most of these microorganisms have the potential to produce a far greater number of natural products than have been isolated previously (Scherlach and Hertweck 2009). It seems that most biosynthetic gene clusters are either "silent" or "cryptic" pathways. Low production rates and large metabolic backgrounds may lead to overlooking respective metabolites as well. Although there is clear correlation between the developmental stages of microbial cultures and the onset of secondary metabolite production, a detailed understanding of the mechanisms underlying this global control is lacking. The involvement of chemical or environmental signals necessary for triggering these pathways has been proposed but only scarcely proofed. Only recently the first signaling cascade for the linkage of nutrient stress to antibiotic production was identified in a Streptomyces strain (Rigali et al. 2008). However, several strategies for the triggering of silent gene clusters have been developed. Some strategies rely on statistical chances, as the one strain-many compounds (OSMAC) approach, where new compounds are derived by application of as many different cultivation conditions as possible. The parameters include media composition, aeration rate, type of culture vessel, and the addition of enzyme inhibitors (Bode et al. 2002). Alternatively, in organisms amendable to genetic manipulation, transcription regulators and promoters are changed by molecular techniques. Other strategies take advantage of genome mining, from the utilization of bioinformatically predicted physicochemical properties to methods that exploit a probable interaction of microorganisms (Chiang et al. 2011).

The knowledge of natural conditions, which leads to the expression of a given gene cluster, is still limited but would offer a good basis for the development of new cultivation techniques. One attempt is the simulation of the natural habitat by cocultivation of microorganisms from the same habitat. Another possibility is the utilization of compounds, acting as quorum-sensing molecules. These simulation strategies will help discover new natural products. As a second benefit, these experiments will enlarge our knowledge on microbial communication. Genomic mining seeks to exploit the hidden potential of biosynthetic pathways. This fascinating, interdisciplinary research field began to form with the development of new sequencing technologies at the turn of the millennium. They offer fast and cheap high throughput sequencing approaches. The impact of genomics and proteomics on the biotechnological exploitation of marine microbiota has hardly been felt yet. The presence of many new gene families from uncultured and highly diverse microbial populations also represents a rich source of new metabolites and enzymes for drug development. Metagenomic strategies used as culture-independent methods, such as isolating and analyzing PKS (Polyketide synthases) gene clusters, have recently provided first insights into the chemical potential of communities of sponge-associated bacteria. These studies revealed two evolutionarily distinct, unusual PKS types that are commonly found in sponge metagenomes and were shown to be of bacterial origin (Teta et al. 2010). While metagenomic approaches are useful for exploiting the biochemistry of microbial communities, they are unable to access the metabolic capabilities of specific microorganisms within these communities. It is also quite likely that many novel genes (coding for enzymes, natural product biosynthesis gene clusters or others) from rare microbes in complex communities are poorly or not represented in metagenomic libraries (Kennedy et al. 2008). Single cell genomics using multiple displacement amplification will complete the technological bundle and allow the study of the entire biochemical potential of single uncultured microbes from complex microbial communities (Woyke et al. 2009). This new approach has great potential for the discovery of novel enzymes and natural products from cyanobacteria, as it affords easier access to rare microbiota. While the standard metagenomic approach may indicate the presence of genes within a given environment, it does not answer the question: Which of these genes are active within the environment? To overcome this particular problem, metatranscriptomics-based approaches need to be employed to study microbial populations, in which only transcriptionally active genes are accessed (Gilbert and Dupont 2011). An alternative strategy to access the metagenome is to directly analyze the proteins within the environment using metaproteomics (Schweder et al. 2008). However, these new and promising methods have not yet been applied on biosynthesis of secondary metabolites in cyanobacteria but are quite promising in the development of drugs from them.

5.3   Future Prospects

Cyanobacteria in general and extremophilic cyanobacteria, in particular, are excellent candidates of structurally diverse bioactive metabolites having the potential for novel drug development. They have already showed their capability as many of these valuable metabolites are in different stages of clinical trial. Significant biological activities such as antitumor, antiviral, antibacterial, antifungal, antimalarial, antimycotics, antiproliferative, cytotoxicity, immunosuppressive agents and multidrug resistance reversers and their associated cellular targets have been reported for several cyanobacterial metabolites. The range of cellular targets is diverse and they include cytoskeletal structures, for example microtubules and actin filaments, as well as enzymes, such as proteasome and histone deacetylases. It is this attribute that makes cyanobacterial compounds highly attractive as potential drug leads and molecular probes. For instance, ~24% of marine natural products commercially available for biomedical research are of cyanobacterial origin (Tan 2013). When the predicted biosynthetic origins of 20 marine-derived molecules, as approved drugs or in clinical trials, were taken into consideration 20% originated from marine cyanobacteria (Tan 2013). An impressive 533 natural products have been reported from only marine strains of cyanobacteria (Marin lit). However 90% of all of these molecules are attributed to only five cyanobacterial genus (Marin lit). There is a vast diversity of extremophilic cyanobacteria in various extreme habitats, which is not yet explored. Many promising metabolites from them could not be developed into potential therapeutic agents owing to supply problems. New methods are ardently needed for the efficient culture of these organisms, taking into account their unique nutrient and physical requirements and, in some cases, growth at low-cell densities.

With our increased understanding of genetics and biosynthesis of natural products, the regulation of natural product biosynthesis can be optimized (Unsin et al. 2013). The biosynthesis of natural products themselves can also be manipulated to yield new derivatives with possibly superior qualities and quantities. In addition to identifying new natural products, genome mining would certainly have an impact on the understanding and manipulation of natural product production. Synthetic drugs typically are the result of numerous structural modifications over the course of an extensive drug discovery program, whereas a natural product can go straight from "hit" to drug. Cyanobacterial natural products are notable not only for their potential therapeutic activities but also for the fact that they frequently have the desirable pharmacokinetic properties required for clinical development. In order for drug discovery from cyanobacteria to continue to be successful, new and innovative approaches are required. By applying these new approaches in a systematic manner to natural product drug discovery, it might be possible to increase the current efficiency in identifying and developing new drugs from cyanobacteria. The genome sequencing projects have opened our eyes to the overlooked biosynthetic potential and metabolic diversity of cyanobacteria. While traditional approaches have been successful at identifying many useful therapeutic agents from these organisms, new strategies are needed in order to exploit their true biosynthetic potential. Several genomics-inspired strategies have been successful in unveiling new metabolites that were overlooked under standard fermentation and detection conditions. In addition, genome sequences have given us valuable insight into genetically engineering biosynthesis gene clusters that remain silent or are poorly expressed in the absence of a specific trigger. As more genome sequences are becoming available, we are noticing the emergence of underexplored or neglected organisms as alternative resources for new therapeutic agents (Winter et al 2011). The modularity and colinearity of the cyanobacterial PKS-NRPS (Polyketide synthases- nonribosomal peptide synthetases) gene clusters are attractive features for the heterologous expression of natural products and genetic manipulation for combinatorial biosynthesis of new hybrid chemical entities. In order to achieve success in genetic applications to produce compounds, it is important to have a better understanding at the molecular level of these gene clusters and the underlying enzymology associated with the biosynthetic pathways. For instance, a recent study by Copp and Neilan (2006) revealed unique function-based phosphopantetheinyl transferases in cyanobacteria through phylogenetic analysis. Such information would have important implications in the heterologous production of cyanobacterial secondary metabolites. However, more research is needed in practical utility of this technique to develop drug candidates. Culture-independent methods using informatic analyses of metagenomes can be used to identify unique secondary metabolite gene clusters, and these genes can subsequently be synthesized and introduced into expression hosts (Gerwick, and Fenner 2013). However, in principle as a generic solution, it has been difficult to enact in more than a few model cases, and technological knowledge of the pathways and their operation is still fragmentary and in need of additional research.

In recent years the process of extract preparation and bioassay-guided fractionation are increasingly automated with partially or fully purified materials being used upfront to avoid crude extracts altogether. The past few years have witnessed major developments in fermentation optimization, purification, dereplication, and structure elucidation of natural products, thus enabling much faster access to sufficient amounts of pure compounds (Lam 2007). The recent development in the hyphenated techniques, which combine separation technologies such as high-pressure liquid chromatography (HPLC) and solid phase extraction (SPE) with nuclear magnetic resonance (NMR), and mass spectrometry (MS), has had a substantial impact in shortening the timeline for dereplication, isolation, and structure elucidation of the natural products present in the crude extracts (Lam 2007). Technological advances are also being made in various sectors to improve access to minor metabolites, such as the increasingly widespread use of NMR microcryogenic and capillary flow probes, biological assays in increasingly smaller volumes such as in 384- and 1534-well plate formats, and, perhaps most powerfully, enhancements to the methods, as well as informatics associated with mass spectrometry so that the supply issue can be minimized. The semi-synthetic approach to drug development can generate analogs by modifying the existing functional groups of natural products. Although this approach leads to less diversity in terms of structural variety, it has certainly yielded many exciting lead natural product molecules with improved properties over the parent compounds.

The use of natural products as templates for constructing biologically relevant chemical libraries is a logical extension of the classical combinatorial library synthesis protocol. Numerous libraries incorporating a natural product motif have been published recently; this shows that novel biologically active analogs can often be discovered by this process (Tan 2005). After identifying genuine natural product leads, applying new organic synthetic methodologies, bio-transformation, combinatorial biosynthesis, and combinations of these techniques for the modification of natural product leads would generate a large number of novel, structurally diverse analogs that can be screened for improved properties as a novel drug candidate.

References

Apt KE, Behrens PW (1999) Commercial development in microalgal biotechnology. J Phycol 35:215–226

Bhatnagar I, Kim SK (2010) Immense essence of excellence: marine microbial bioactive compounds. Mar Drugs 8:2673–2701

Blokhin AV, Yoo HD, Geralds RS, Nagle DG, Gerwick WH, Hamel E (1995) Characterization of the interaction of the marine cyanobacterial natural product curacin A with the colchicine site of tubulin and initial structure-activity studies with analogs. Mol Pharmacol 48:523–531

Bode HB, Bethe B, Hofs R, Zeeck A (2002) Big effects from small changes: possible ways to explore nature's chemical diversity. Chem biochem 3:619–27

Chiang YM, Chang SL, Oakley BR, Wang CCC (2011) Recent advances in awakening silent biosynthetic gene clusters and linking orphan clusters to natural products in microorganisms. Curr Opin Chem Biol 15:137–143

Copp JN, Neilan BA (2006) The phosphopantetheinyl transferase superfamily: phylogenetic analysis and functional implications in cyanobacteria. Appl Environ Microbiol 72:2298–2305

D'Agostino G, Del Campo J, Mellado B, Izquierdo MA, Minarik T, Cirri L, Marini J, Perez-Gracia L, Scambia G (2006) A multicenter phase II study of the cryptophycin analog LY355703 in patients with platinum-resistant ovarian cancer. Int J Gynecol Cancer 16:71–76

Dixit RB, Suseela MR (2013) Cyanobacteria: potential candidates for drug discovery. A Van Leeuw J Microb 103:947–961

Ebbinghaus S, Hersh E, Cunningham CC, O'Day S, McDermott D, Stephenson J, Richards DA, Eckardt J, Haider OL, Hammon LA (2004) Phase II study of synthadotin (SYN-D; ILX651) administered daily for 5 consecutive days once every 3 weeks (qdx5q3w) in patients (Pts) with inoperable locally advanced or metastatic melanoma. In Proc Am Soc Clin Oncol 24:A7530

Gerwick WH, Fenner AM (2013) Drug discovery from marine microbes. Microb Ecol 65:800–806

Gerwick WH, Proteau PJ, Nagle DG, Hamel E, Blokhin A, Slate DL (1994) Structure of curacin A, a novel antimitotic, antiproliferative, and brine shrimp toxic natural product from the marine cyanobacterium *Lyngbya majuscula*. J Org Chem 59:1243–1245

Gilbert JA, Dupont CL (2011) Microbial metagenomics: beyond the genome. Ann Rev Mar Sci 3:347–371

Hayashi O, Katoh T, Okuwaki Y (1994) Enhancement of antibody production in mice by dietary *Spirulina platensis*. J Nutr Sci Vitaminol 40:431–441

Kennedy J, Marchesi JR, Dobson ADW (2008) Marine metagenomics: strategies for the discovery of novel enzymes with biotechnological applications from marine environments. Microb Cell Fact 7:27

Khan Z, Bhadouria P, Bisen PS (2005) Nutritional and therapeutic potential of Spirulina. Curr Pharm Biotechnol 6:373–379

Kim JD, Lee CG (2005) Systemic optimization of microalgae for bioactive compound production. Biotechnol Bioprocess Eng 10:418–424

Kim ZH, Kim SH, Lee HS, Lee CG (2006) Enhanced production of astaxanthin by flashing light using *Haematococcus pluvialis*. Enzyme Microb Technol 39:414–419

Lam KS (2007) New aspects of natural products in drug discovery. Trends Microbiol 15:279–289

Liang J, Moore RE, Moher ED, Munroe JE, Al-awar RS, Hay DA, Varie DL, Zhang TY, Aikins JA, Martinelli MJ, Shih C, Ray JE, Gibson LL, Vasudevan V, Polin L, White K, Kushner J, Simpson C, Pugh S, Corbett TH (2005) Cryptophycins-309, 249 and other cryptophycin analogs: preclinical efficacy studies with mouse and human tumors. Invest New Drugs 23:213–224

Munro MHG, Blunt JW (2009) MarineLit. University of Canterbury http://pubs.rsc.org/marinlit

Márquez B, Verdier-Pinard P, Hamel E, Gerwick WH (1998) Curacin D, an antimitotic agent from the marine Cyanobacterium *Lyngbya majuscula*. Phytochemistry 49:2387–2389

Mayer AMS, Glaser KB, Cuevas C, Jacobs RS, Kem W, Little RD, McIntosh JM, Newman DJ, Potts BC, Shuster DE (2010) The odyssey of marine pharmaceuticals: a current pipeline perspective. Trends Pharmacol Sci 31:255–265

Mita AC, Hammond LA, Bonate PL, Weiss G, McCreery H, Syed S, Garrison M, Chu QS, DeBono JS, Jones CB, Weitman S, Rowinsky EK (2006) Phase I and pharmaco-kinetic study of tasidotin hydrochloride (ILX651), a third-generation dolastatin-15 analogues, administered weekly for 3 weeks every 28 days in patients with advanced solid tumors. Clin Cancer Res 12:5207–5215

Newman DJ, Cragg GM, Snader KM (2003) Natural products as sources of new drugs over the period 1981–2002. J Nat Prod 66:1022–1037

Park HJ, Lee YJ, Ryu HK, Kim MH, Chung HW, Kim WY (2008) A randomized double-blind, placebo-controlled study to establish the effects of *spirulina* in elderly Koreans. Ann Nutr Metab 52:322–328

Patterson GM, Baldwin CL, Bolis CM, Caplan FR, Karuso H, Larsen LK, Levine IA, Moore RE, Nelson CS, Tschappat KD, Tuang GD, Furusawa E, Furusawa S, Norton TR, Raybourne RB (1991) Antineoplastic activity of cultured blue–green algae (Cyanophyta). J Phycol 27:530–536

Qureshi M, Ali R (1996) *Spirulina platensis* exposure enhances macrophage phagocytic function in cats. Immunopharmacol Immunotoxicol 18(3):457–463

Ramaswamy AV, Flatt PM, Edwards DJ, Simmons TL, Han B, Gerwick WH (2006) The secondary metabolites and biosynthetic gene clusters of marine cyanobacteria. Applications in biotechnology. In: Proksch P, Müller WEG (eds) Frontiers in marine biotechnology. Horizon Bioscience, Norfolk, pp 175–224

Rigali S et al (2008) Feast or famine: the global regulator DasR links nutrient stress to antibiotic production by *Streptomyces*. EMBO Rep 9:670–675

Scherlach K, Hertweck C (2009) Triggering cryptic natural product biosynthesis in microorganisms. Org Biomol Chem 7:1753–1760

Schweder T, Markert S, Hecker M (2008) Proteomics of marine bacteria. Electrophoresis 29:2603–2616

Simmons TL, Nogle LM, Media J, Valeriote FA, Mooberry SL, Gerwick WH (2009) Desmethoxymajusculamide C, a cyanobacterial depsipeptide with potent cytotoxicity in both cyclic and ring-opened forms. J Nat Prod 72:1011–1016

Suh IS, Lee CG (2003) Photobioreactor engineering: design and performance. Biotechnol Bioprocess Eng 8:313–321

Tan DS (2005) Diversity-oriented synthesis: exploring the intersections between chemistry and biology. Nat. Chem Biol 1:74–84

Tan LT (2013) Marine cyanobacteria: a prolific source of bioactive natural products as drug leads. In: Kim S-K (ed) Marine microbiology: bioactive compounds and biotechnological applications. Wiley-VCH Verlag GmbH & Co. KGaA, Weinheim, pp 59–81

Teta R, Gurgui M, Helfrich EJ, Künne S, Schneider A, Echten-Deckert V, Mangoni A, Piel J (2010) Genome mining reveals trans-AT polyketide synthase directed antibiotic biosynthesis in the bacterial phylum Bacteroidetes. ChemBioChem 11:2506–2512

Unsin CEM, Rajski SR, Shen B (2013) The role of genetic engineering in natural product-based anticancer drug discovery. In: Koehn FE (ed) Natural products and cancer drug discovery. Springer, New York, pp. 175–191

Uzair B, Tabassum S, Rasheed M, Rehman SF (2012) Exploring marine cyanobacteria for lead compounds of pharmaceutical importance. Sci World J 2012:179782

Watanabe J, Minami M, Kobayashi M (2006) Antitumor activity of TZT-1027 (soblidotin). Anticancer Res 26:1973–1981

Winter JM, Behnken S, Hertweck C (2011) Genomics-inspired discovery of natural products. Curr Opin Chem Biol 15:22–31

Wipf P, Reeves JT, Day BW (2004) Chemistry and biology of curacin A. Curr Pharm Des 10:1417–1437

Woyke T, Xie G, Copeland A, Gonzalez JM, Han C, Kiss H, Saw JH, Senin P, Yang C, Chatterji S, Cheng JF, Eisen JA, Sieracki ME, Stepanauskas R (2009) Assembling the marine metagenome, one cell at a time. PLoS ONE 4:e5299

Yoo HD, Gerwick WH (1995) Curacins B and C, New antimitotic natural products from the marine cyanobacterium *Lyngbya majuscula*. J Nat Prod 58:1961–1965

Yousong D, Seufert H, Zachary QB, David HS (2008) Analysis of the cryptophycin P450 epoxidase reveals substrate tolerance and cooperativity. J Am Chem Soc 130:5492–5498

Index

© Springer International Publishing Switzerland 2015
S. Mandal, J. Rath, *Extremophilic Cyanobacteria For Novel Drug Development*,
SpringerBriefs in Pharmaceutical Science & Drug Development,
DOI 10.1007/978-3-319-12009-6